NASA MARS ROVERS

1997–2013 (Sojourner, Spirit, Opportunity and Curiosity)

DEDICATION
To the men and women of the EDL (Entry, Descent and Landing) teams, without whose expertise and professionalism none of these events would have been successful.

First published in May 2013

A catalogue record for this book is available from the British Library.

ISBN 978 0 85733 370 4

Library of Congress control no. 2012955425

Published by Haynes Publishing,
Sparkford, Yeovil,
Somerset BA22 7JJ, UK.
Tel: 01963 442030 Fax: 01963 440001
Int. tel: +44 1963 442030 Int. fax: +44 1963 440001
E-mail: sales@haynes.co.uk
Website: www.haynes.co.uk

Haynes North America Inc.
861 Lawrence Drive, Newbury Park,
California 91320, USA.

Printed in the USA by Odcombe Press LP,
1299 Bridgestone Parkway, La Vergne, TN 37086.

Acknowledgements

The author would like to acknowledge the help, support and cooperation of countless individuals and numerous organizations in the US space programme, especially those at the Jet Propulsion Laboratory – home of the Mars rovers – and at NASA headquarters. On the inception and birth of this Haynes book, the author thanks Mark Hughes for initiating the project and for his outstanding support throughout, Steve Rendle for working tirelessly toward its design and production, Ian Moores for some of the line drawings and Jonathan Falconer for having been there for advice. Last, but certainly not least, to Ann. She knows why.

Useful contacts

British Interplanetary Society
27/29 South Lambeth Road, London SW8 1SZ, UK
Tel 020 7735 3160
www.bis-spaceflight.com
Signatory to the International Astronautical Federation, the BIS was formed in 1933 with Arthur C. Clarke an early member. The Society is open to all and has two publications available on subscription. Holds regular meetings with lectures and has a library open to members.

National Space Centre
Exploration Drive, Leicester LE4 5NS, UK
Tel 0845 605 2001
www.spacecenter.co.uk
Provides an all-round educational experience in many separate aspects of space research and exploration.

The Science Museum
Exhibition Rd, South Kensington SW7 2DD, UK
Tel 0870 870 4868
www.sciencemuseum.org.uk
Contains a full space gallery with many relevant exhibits including a Shuttle model and artifacts from the space programmes of Europe, the UK, the US, Russia and China.

International Space University
Parc d'Innovation, 1 rue Jean-Dominique Cassini, 67400 Illkirch-Graffenstaden, France
Tel 0033 3 88 65 54 30
www.isunet.edu
Provides graduate-level educational resources, courses and qualifications for the aspiring space professional, having graduated more than 3,000 students from 100 countries.

National Air & Space Museum
6th & Independence Avenue SW,
Washington DC 20560, USA
Tel 001 202 633 1000
www.nasm.si.edu
The world's largest aerospace museum with numerous galleries and exhibits covering the space programme.

www.jpl.nasa.gov
Access to comprehensive information about Mars and NASA planetary exploration programmes.

www.nasaspaceflight.com
General space news, forums and a subscription section with insider blogs and contributions from space professionals through dedicated forums.

www.spaceflightnow.com
Space news and detailed information on Mars exploration and all aspects of the space programme.

www.nasa.gov
NASA's home page for a wide range of web pages dedicated to all aspects of space flight and with access to detailed Mars information.

www.esa.int
The webs site of the European Space Agency with access to information and images of the European contribution to the exploration of Mars.

www.ukspaceagency.bis.gov.uk
The British Space Agency with detailed information about the 68,000 jobs and numerous UK businesses aligned with general space activities and its work within ESA on Mars exploration.

COVER CUTAWAY: *Ian Moores.*

NASA MARS ROVERS

1997–2013 (Sojourner, Spirit, Opportunity and Curiosity)

Owners' Workshop Manual

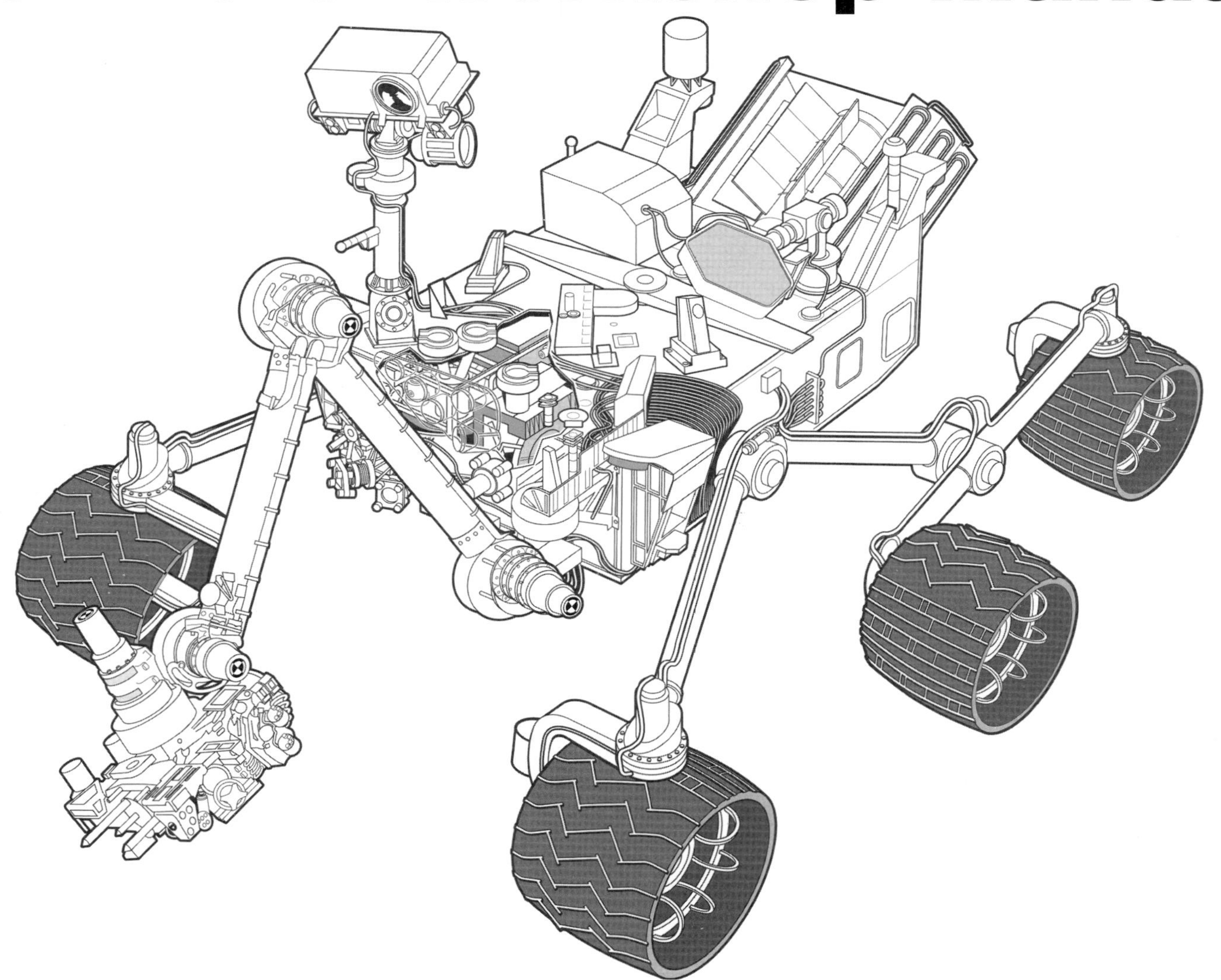

An insight into the technology, history and development of NASA's Mars exploration roving vehicles

David Baker

Contents

OPPOSITE NASA's flagship Mars rover, Curiosity, takes a self-portrait on the surface of Gale Crater, as it begins many years of work searching for places life could have emerged in the distant past. Only by building on the success of three previous rovers, could scientists have developed the biggest wheeled vehicle ever to roll across the surface of another world. *(JPL)*

Introduction

BELOW **Mars is an enigmatic planet, similar to Earth in many respects yet so very different in others. It has giant canyon lands and several massive volcanoes, the largest being three times the height of Mount Everest.** *(NASA)*

The sands of Barsoom are strewn with the wreckage of dashed expectations and the forlorn dreams of science fiction writers yearning to tread the red dust of Mars. For centuries men and women have looked toward the heavens and wondered at the strange, red-coloured star that moves more rapidly across the sky than any other pinpoint of light, except for Mercury, Venus, Jupiter and Saturn. It appears to move in an irregular pattern, defying logic and implying a defined purpose. So it was that in their imaginations Mars was perceived as a god-like being, its red colour being associated with war and strife. A presumption conveyed by writers in fiction and in poetry.

A concoction of fiction writer Edgar Rice Burroughs during the early years of the 20th century, Barsoom was the name given by that writer through the language he imagined for a race of Earth-like beings fighting for diminishing resources on a dying planet. Their struggles became an analogue for human suffering, a conveyance used by Burroughs to give his Martians an identity not so far removed from our own: little people struggling against the titanic forces of a world in change, a place threatened by planetary changes they were powerless to halt.

What scientists saw was less imaginative and devoid of fantasy. Through telescopes it had already become apparent that because it rotates at an angle of 24.8° to the plane of its orbit about the Sun, Mars has Earth-like seasons. Polar ice caps could be seen, broad and extensive in winter, melting in summer months when the surface of Mars darkens towards the equator. Observations from Earth indicated it had an atmosphere thinner than Earth's, probably as low as 10–30% as dense as that of our own planet. From this it was surmised that daytime temperatures at the equator could be around 25°C, plunging to sub-zero at night due to the thin atmosphere.

Around 1878 the Milanese astronomer Giovanni Schiaparelli observed what he believed to be fine lines etched across the surface, markings he called 'canals', and large dark areas he called 'oceans'. Smaller dark patches he referred to as 'lakes', while yellowish regions he called 'continents'. Excited by the detailed observations, astronomers around the world trained their telescopes on Mars and found the thin lines linked these newfound seas and lakes, implying some form of irrigation.

Stimulated by this, in 1895 the US astronomer Percival Lowell took a great leap in imagination and proposed that the canals were artificial, created by intelligent Martians to move water from the melting polar caps to grow crops and plants in arid desert-like places. Reinforcement of the theory appeared when Schiaparelli and Lowell noticed colour changes on Mars suggesting broad expanses of crops spreading across the surface in summer months, changing back again in the autumn, presumably after the Martians had harvested their crops.

Fictional depictions of these Martians soon aroused popular interest, and speculation fuelled by pseudo-science ran rife, exploited by the new media of comics and radio – so much so that in October 1938 Orson Welles purportedly spread panic in American homes with his realistic rendition of H.G. Wells' famous science fiction novel *The War of the Worlds*. Confused by its news-report format, some listeners allegedly believed it to be a live report of a Martian invasion of Earth, but the extent to which it shocked Americans is highly questionable and due largely to the Orson Welles' publicity machine.

Nevertheless, not until July 1965 would Earthlings receive the first close-up pictures of the Red Planet and get a better idea as to whether or not there really were Martians living on the surface.

The possibility of sending spacecraft to Mars looked suddenly more likely after the launch of Sputnik 1 on 4 October 1957, and artificial satellites launched by the Soviet Union and the United States over the following months encouraged active planning to achieve that. The practicality of sending man-made objects into space opened a new era in exploration, but it was happening in a slightly different way to that which had been expected. When the Second World War ended in 1945 the German V-2 had demonstrated that rockets could be built capable of carrying scientific instruments into space. Bigger and more powerful rockets could place satellites in Earth orbit and further development with upper stages would accelerate spacecraft to escape velocity and voyage on to the planets.

Most people expected space to be explored by humans. Artificial robots were thought by the average person as even more unlikely than the prospect of space travel. But rapid advances in radio communication, radar and electronics during and immediately after the Second World War opened the prospect of first sending machines instead of men and women. While research into high-speed aircraft pushing ever closer to the edge of space was progressing, the development of rockets and ballistic missiles meant that machines could get there faster and at much less cost.

When space became a political race in the Cold War between the Soviet Union and the United States, unmanned spacecraft were set to pioneer a path across the new frontier. No longer were the giant rocket ships prophesied by V-2 rocket engineer Wernher von Braun a logical way to go. Smaller craft would go first, carrying scientific instruments and radio equipment to transmit their information.

By the late 1950s engineers were planning ways to send spacecraft to the Red Planet – and to traverse its surface when they got there.

David Baker
February 2013

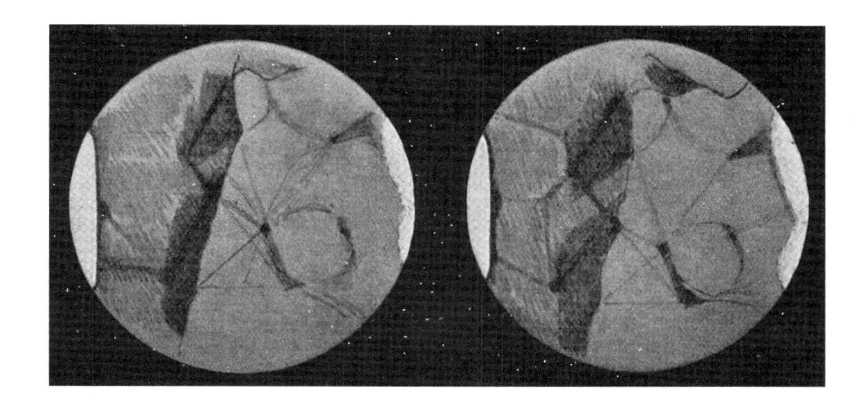

ABOVE In the United States, Percival Lowell, later made famous by his discovery of the planet Pluto, imaged artificial canals across the surface of Mars. This did little to suppress fears that intelligent beings could be plotting attacks on Earth! *(Getty Images)*

DAILY NEWS FINAL

NEW YORK'S PICTURE NEWSPAPER

FAKE RADIO 'WAR' STIRS TERROR THROUGH U.S.

"War" Victim

"I Didn't Know"

LEFT Fictional Mars-attacks migrated from irrational fear to hysteria, exemplified by Orson Welles' play based on H G Wells' *War of the Worlds*. *(Getty Images)*

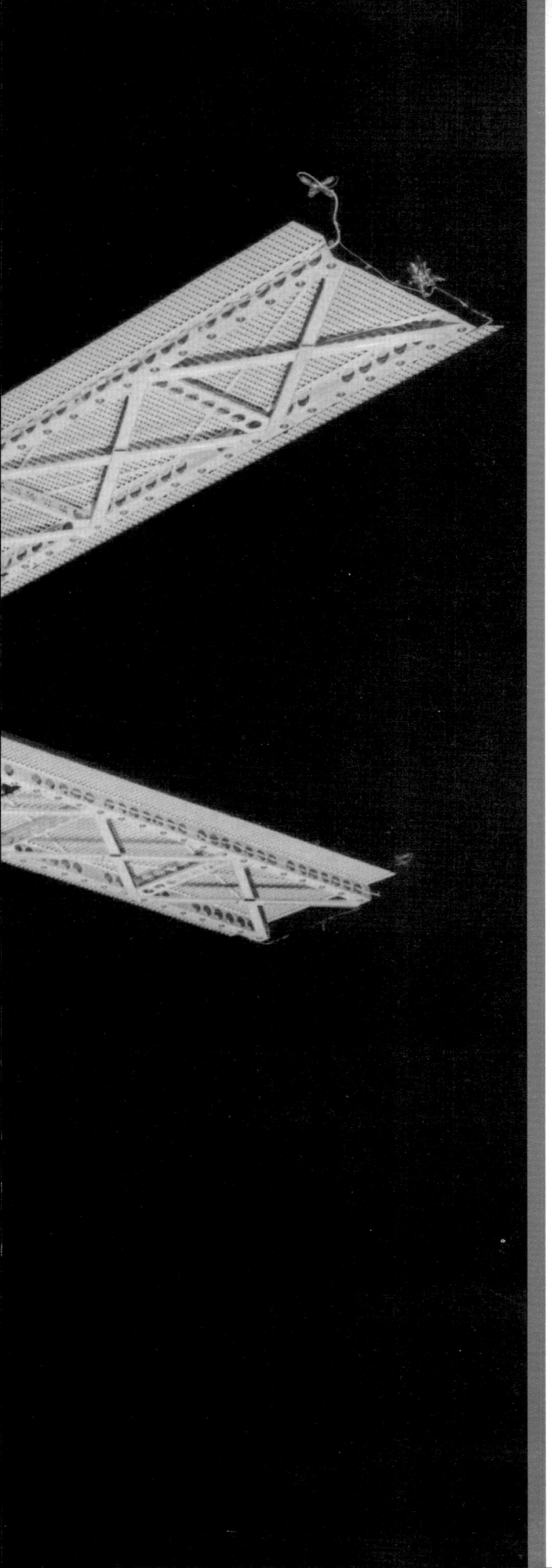

Chapter One

Target Mars

The dawn of the Space Age realised dreams of reaching other worlds and a prime destination was Mars – known as the Red Planet for its distinctly red-coloured appearance in the night sky. Was it populated by alien civilisation bent on destruction of Earthlings as imagined by science fiction writers – or was it a world like our own with seas and lush vegetation? Only our robotic ambassadors could tell us.

OPPOSITE Mariner 9 carried a scan platform with science instruments including two telescopic cameras which effectively mapped the entire planet and revealed a much more fascinating world than had previously been thought. *(NASA)*

Mars has been enigmatic ever since the first humans gazed at this reddish spot in the night sky and Roman soldiers from the Apennine Way to Hadrian's Wall linked it to the god of war. Inspired by mythology, science fiction writers down the centuries have used Mars as a home for aliens, beings invariably bent on dominating Earth, so there was no surprise that when the space age began in the late 1950s, Mars was an early target for exploration.

Mars is an outer planet and goes around the Sun outside the orbit of the Earth. While Earth takes 365 days to go round the Sun, Mars makes one complete orbit every 687 days, almost 23 Earth months. This is not only because it is further from the Sun and has more distance to travel in its orbit, but also because its orbital speed is less and the planet takes longer to travel a given distance in its orbit.

Mars is smaller than the Earth but larger than the Moon, its surface area being approximately the same as all the dry land on Earth. However, Mars spins on its polar axis a little more slowly and a 'day' on Mars is 40 minutes longer than an Earth day. But Mars has an atmosphere and a solid surface with a history going back to the origin of the solar system more than 4.5 billion years ago, when it may have been much more like Earth than it is at present.

BELOW Seen here undergoing pre-flight tests at the Jet Propulsion Laboratory (JPL) before its launch on 5 November 1964, NASA's Mariner 4 was the first US spacecraft sent to Mars. *(JPL)*

To get to Mars, Earth must be aligned with the planet so that the outbound arc of the flight path will carry the spacecraft to the point in space where Mars will be when it arrives. Scientists have known about the orbit of Mars for a long time and by the dawn of the space age had calculated the best way to reach the Red Planet. Flight paths, or trajectories, can be one of two basic types on what are known as 'Hohmann transfer orbits', named after the German engineer Walter Hohmann who in 1925 calculated the most energy-efficient ways of reaching other planets.

Type I trajectories are those which require the spacecraft to travel less than 180° around the Sun to reach their destination; Type II trajectories take more than 180°. But in each case the precise alignment of the Earth and the target planet defines what engineers call the 'launch window'. In the case of Mars this occurs once every 26 months.

Designing for Mars

In 1964 NASA launched a 575lb (261kg) Mariner spacecraft to Mars with a camera for taking pictures as it flew past the planet. Photographs were to be taken in groups of two with a small gap between each pair; some overlap was expected, the amount depending on the precise fly-by distance from the surface. At a transmission rate of 8.3 bits/sec, it would take 8hr 20min to send a complete picture back to Earth. At most, Mariner 4 would shoot 22 pictures and it would take 211hrs to get all those reassembled on Earth.

One vital element of Mariner and all future planetary missions to any destination in the solar system was the development of the NASA Deep Space Network. Without the DSN, no signals would send back to Earth the vital information gathered along the way and at the planetary target. Three large dish antennas, located at three equally spaced places around the circumference of the planet, were constructed so that however the Earth was positioned at least one would be within direct sight of the spacecraft.

The first, built at Goldstone, California, was an 85ft (26m) dish antenna for tracking early Moon probes. Designated DSIF

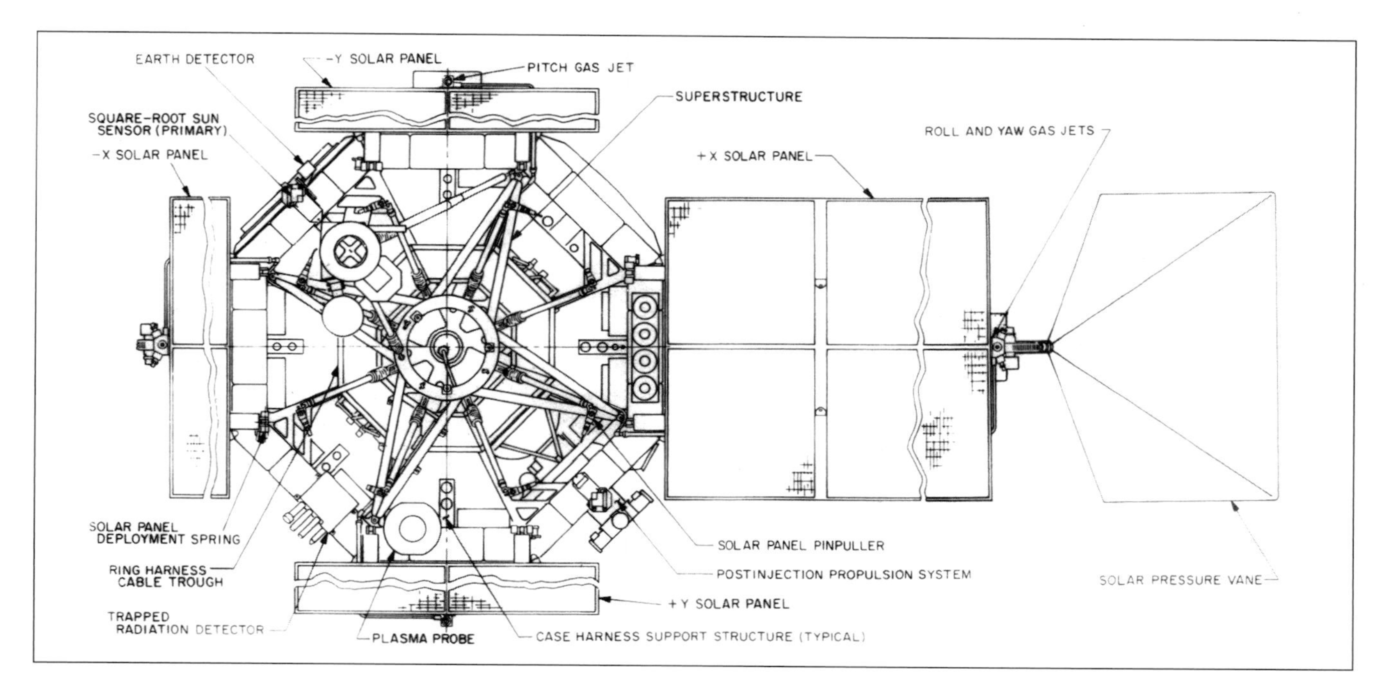

ABOVE The basic frame for the 575lb Mariner 4 comprised a 30lb eight-sided magnesium frame, 18in high and 54.5in across, incorporating seven electronic compartments. Electrical power was provided by four solar array wings each 71.4in long and 35.5in wide supporting 28,224 cells across an area of 70.4ft^2 providing 310–640W at Earth and Mars distances respectively. *(David Baker)*

(Deep Space Instrumentation Facility) 11, it remained operational until 1981. By 1960 work had begun at Woomera, Australia, and Johannesburg, South Africa, erecting the other two 85ft dishes. Woomera (DSIF 41) would continue to operate until replaced by a site at Canberra in 1971. Johannesburg (DSIF 51) was ready by July 1961, by which time a second 85ft dish had been added at Goldstone.

By 1966 Goldstone had the additional use of a new 210ft (64m) dish antenna through which it could process signals at a much faster rate, a similar dish antenna being added at Canberra in 1973. By the early 1970s the political situation in South Africa led the US government to relocate the tracking job for that longitude to Madrid, Spain, where a 210ft antenna was also built in 1973. By this date, each of the three sites had two networks of 85ft antennas in addition. Long before these events, the DSIF had been renamed the Deep Space Network – which said it all!

Mariner 4 was launched on 28 November 1964 and took almost eight months to reach Mars. On 14 July 1965 the picture recording sequence began with Mariner 4 10,500 miles (16,900km) from the surface and closing. By the time picture number 17 was taken the spacecraft had reached its point of closest approach (6,118 miles), passing high across the surface at more than 11,000mph (17,710km/h). The remaining four pictures were taken as the spacecraft overtook Mars and began to increase its distance from the surface. It took the cameras 26 minutes to record the 21 pictures on tape, but not until later would they be transmitted to the Earth.

About 75 minutes after closest encounter the radio signal was lost as Mariner 4 swept past Mars on the far side from Earth, contact being re-established when it reappeared 54 minutes later. By careful study of the way the side signal 'died' as the radio waves passed through the atmosphere, a process known as 'occultation', scientists were greatly surprised to discover that the atmosphere appeared to have a density less than 0.5% that of the Earth, a lot less than the 10–30% believed by most astronomers; 4–6 millibars compared with around 1,000 millibars at the surface of the Earth. Moreover, it appeared the atmosphere was almost all carbon dioxide instead of 60–95% nitrogen as previously believed.

Approximately 13 hours after the first picture had been taken reassembly of the elements began on Earth, and as the tape recorders whirred on inside the spacecraft, data began to flow – at a mere 8.3 bits/sec – to the big dish antenna on Earth. Painfully slowly, the pictures revealed the surface of Mars for the first time. More than nine days and six hours later, the final complete picture, and 200 lines of another,

LEFT Mariner 4 started taking close-up pictures of Mars on 14 July, passing just 6,118 miles away from the planet's surface and sending 21 complete pictures and part of another. Due to the slow 8.3 bits/sec transmission rate it took more than nine days to get all the images reconstructed on Earth. The images revealed a crater-ridden surface more Moonlike than that expected by many scientists. *(JPL)*

had been received. What they revealed was totally unexpected, a surface pockmarked with Moon-like craters and a continuously undulating and featureless landscape.

Displaying a barren surface densely populated with craters and an atmosphere so thin no water could exist in liquid form at the surface, these results came as a great surprise to scientists. To others the results were a shock, and to those who had believed in Martians it was devastating. Somewhere in the middle, politicians were wondering why anyone would want to go there, which was not what NASA wanted to hear! As Mariner 4 sped on in its eternal orbit of the Sun, it left behind a world now very different in human perception than the one it had been for scientists and science fiction writers alike on faraway Earth. Yet the real exploration of Mars and the astounding discoveries yet to be made were still in the future.

Returning to the Red Planet

NASA returned to Mars in 1969 when Mariners 6 and 7 flew past the planet and took many more pictures. Photographs from Earth-based telescopes could show objects no smaller than 100 miles, whereas the Mariner 4 pictures had shown objects down to about

LEFT NASA built a global communications system, involving three main stations around the world so that wherever the surface was on the planet's rotation at least one Deep Space Network antenna would be in contact with spacecraft on their way to other worlds. Eventually all stations would have 210ft dish antennas like this one at Goldstone, California. Others are located in Spain and Australia. *(JPL)*

10,000ft (approximately 3,000m). The high-resolution cameras on the 1969 spacecraft, however, would resolve objects down to about 900ft (274m). Each image would contain 3.9 million bits compared to the 240,000 bits/image from Mariner 4. The data transmission rate would be at least 270 bits/sec compared to 8.3 bits/sec from Mariner 4. An experimental data rate of 16.2 Kbits/sec was also possible, depending upon the performance of the spacecraft. New to the Mariners was a programmable sequencer with a 128-word core memory that allowed updates and modifications to onboard programmes through ground command.

Mariner 6 was launched on 24 February 1969, followed by Mariner 7 on 27 March 1969. On 31 July, Mariner 6 passed Mars at a distance of 2,131 miles (3,431km) followed on 2 August by Mariner 7 at a height of 2,130 miles (3,430km). Each spacecraft carried single 50mm wide-angle and 508mm narrow-angle cameras and both were programmed to take a large number of pictures during the approach phase as well as during the high-speed flypast. While Mariner 4's 21 pictures covered about 1% of the planet, Mariner 6 and 7 took a total of 1,300 pictures, photographing 20% of the surface. Because the surface area of Mars is roughly the same as the area of dry land on Earth, as a percentage of the whole planet the view obtained by Mariner 4 was about equal to the area of Iran, while that returned from Mariners 6 and 7 was about equal to the area of the United States and Australia combined!

A long, lingering look

By the time scientists on Earth had assimilated the results from Mariners 6 and 7, and found a more interesting planet than had at first been thought from the Mariner 4 pictures, the blueprint for exploration of the Red Planet was firm. What scientists wanted was a long, lingering look at the planet – what they called the 'temporal' view, orbiting Mars for many months photographing the seasonal changes over a lengthy period of time. The next two spacecraft were based on previous Mariners but, at 2,150lb (975kg), they were much heavier due to the propellant they needed to fire a main rocket motor to slow

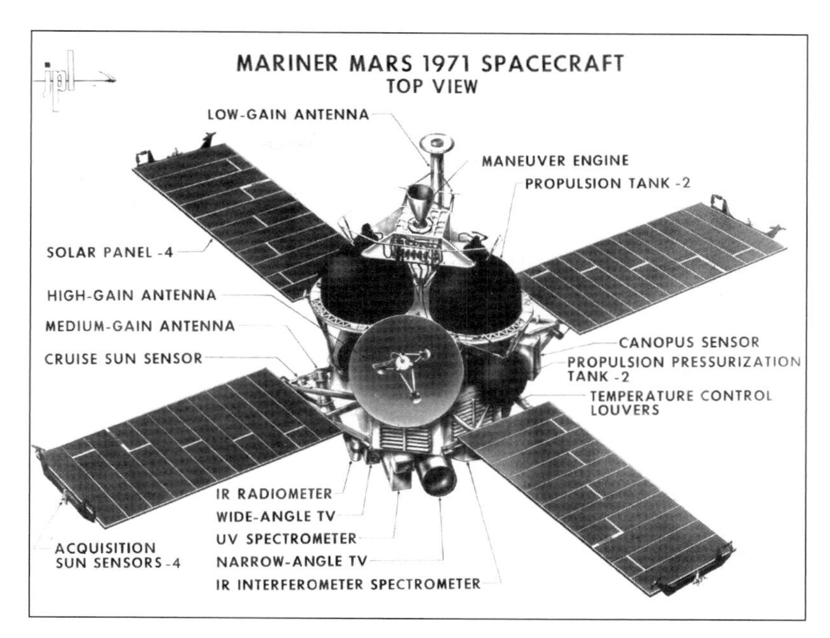

ABOVE After two successful repeats of the Mariner 4 fly-by with Mariners 6 and 7 in 1969, on 30 May 1971 Mariner 9 was sent to enter Mars orbit and spend many months surveying the planet in greater detail. It carried a 300lb thrust rocket motor and two propellant tanks with 1,050lb of hydrazine and nitrogen tetroxide propellants supporting a 15-minute retro-burn to enter an orbit around Mars, slowing the spacecraft by 3,600mph, on 13 November 1971. *(David Baker)*

BELOW To achieve the correct trajectory around Mars, Mariner 9 had to arrive at the planet at precisely the right time, and the correct location where the planet would be in its path around the Sun within a hypothetical box of 435 x 435 miles, far greater accuracy than had been demanded by other planetary missions to Mars and Venus, shown here. *(David Baker)*

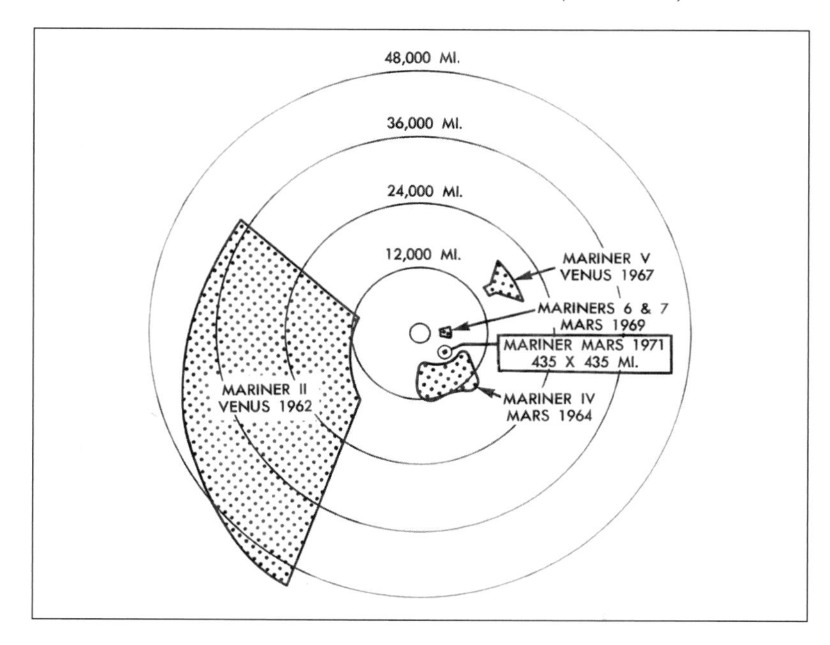

LEFT The early 1970s saw a revolution in solid-stage electronics and increasingly more powerful computer systems for running spacecraft equipment. Built by Hamilton Standard, this onboard Central Computer & Sequencer System was tasked with many more functions than Mariner 4 could handle a mere seven years earlier. *(Hamilton Standard)*

down and allow the gravitational field of Mars to capture them as artificial satellites.

Mariner 8 was launched from Cape Canaveral on 8 May 1971, but a failure in the guidance system in the upper stage resulted in its destruction just six minutes after lift-off; but Mariner 9 was successfully sent on its way on 30 May. Reaching an orbit around Mars on 13 November 1971, Mariner 9 returned 7,329 images of Mars covering 85% of the surface, showing features as small as approximately

RIGHT Tape recorders were the only means of storing data, and with its powerful telescopic images Mariner 9 brought unusually high demands to the data storage system and its associated electronics, returning a total of 7,329 pictures and 54 million bits of data, 27 times the quantity of data previously transmitted by Mariners 4, 6 and 7. *(Hamilton Standard)*

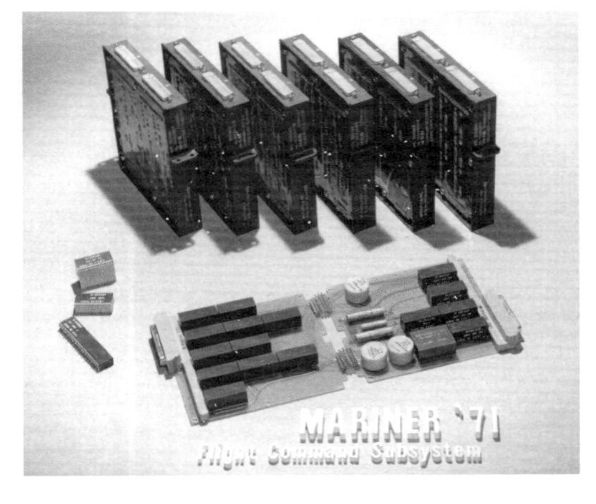

BELOW The Mariner 9 Flight Command Subsystem provided the articulation for commands originating within the CC&SS destined for activation. The FCS handled spacecraft control functions for keeping it operating ('housekeeping' chores) so that the science instruments (the payload) could do their job. *(Hamilton Standard)*

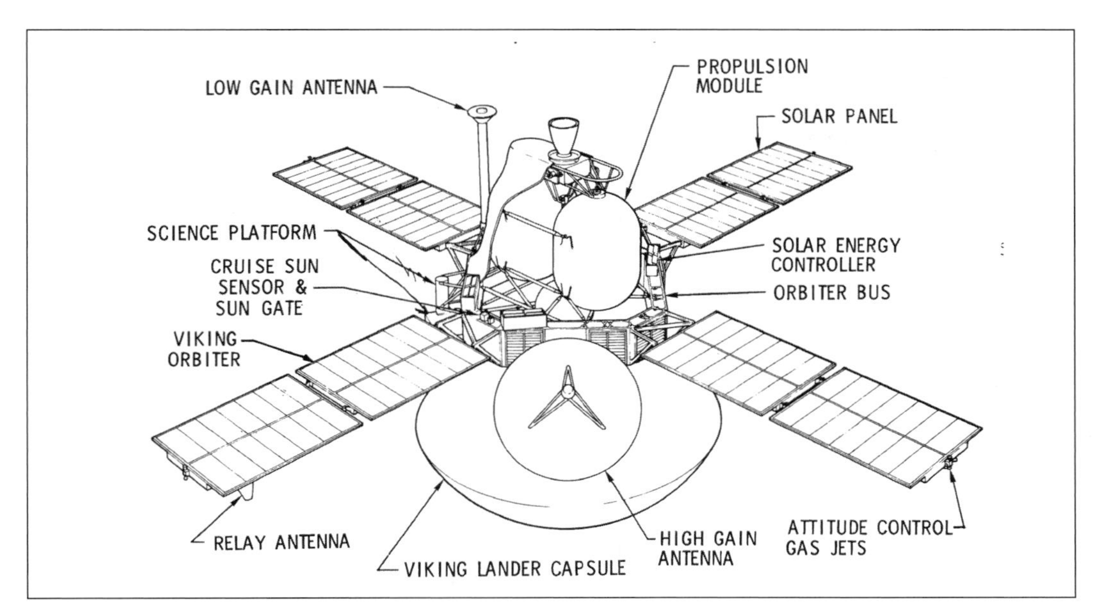

LEFT The first NASA project to land a spacecraft on Mars, including orbiter and lander, Viking weighed a total 7,746lb at launch compared with Mariner 9, which weighed 2,150lb. *(David Baker)*

330ft (100m), and continued to take pictures until 27 October 1972. In all, Mariner 9 transmitted to Earth 54 million bits (binary digits) of information, returning 27 times the quantity of data sent by Mariners 4, 6 and 7.

Looking for life

Between 1964 and 1972 four Mariner spacecraft had transformed scientists' view of the Red Planet. A world once thought to be populated by alien beings, in the imagination of some science fiction writers a warlike force bent on conquering Earth, the planet was now revealed as a dry, dusty place seemingly devoid of life. But the 21 disappointing pictures showing craters and a barren surface transmitted back to Earth by Mariner 4 had given way to a new and exciting vista displayed in the 7,529 pictures transmitted by Mariners 6, 7 and 9 – particularly Mariner 9, which opened entirely new possibilities about the geology and the climate of a world now very different to what it had once been.

After the dust storm that obscured the surface when it arrived had abated, Mariner 9

LEFT Popular space scientist and TV personality Carl Sagan gives scale to a full-size model of the Viking lander. The Viking 1 landing site would be named the Carl Sagan Memorial Station. *(NASA)*

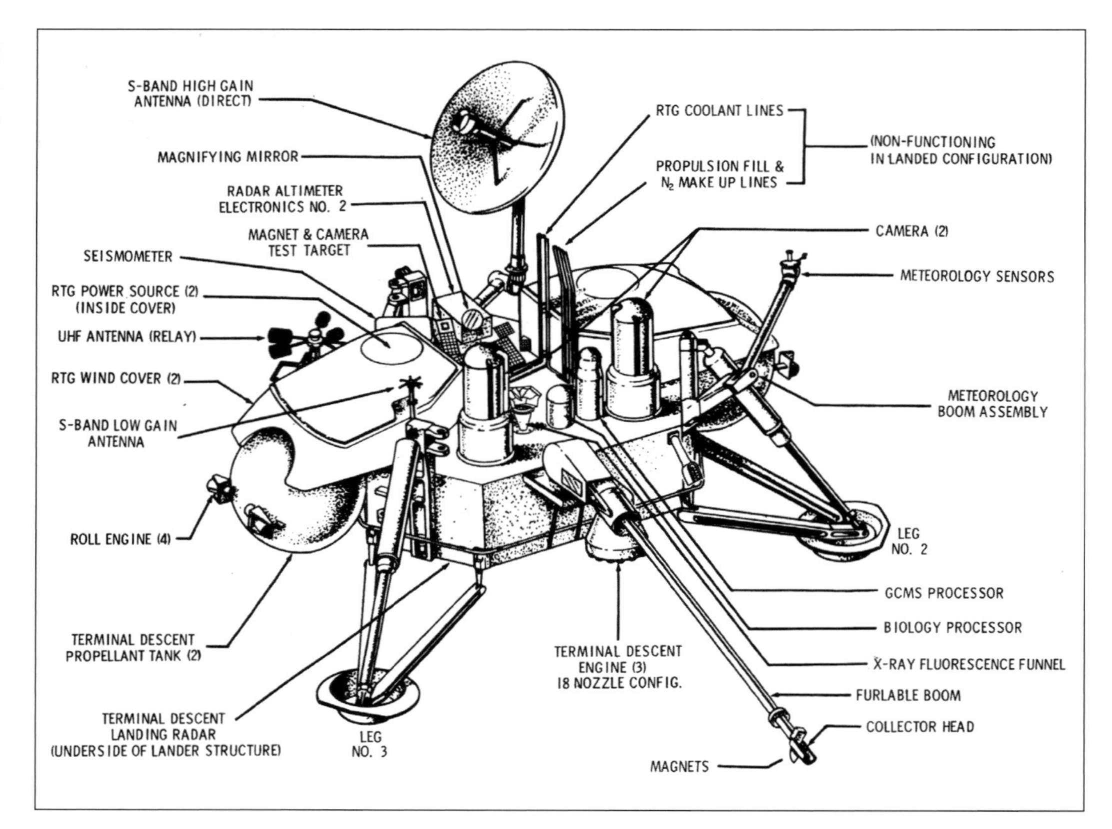

RIGHT **Powered by a radioisotope thermoelectric generator in which heat from plutonium-238 is converted into electrical energy, Viking was designed to last many months in the hostile environment of Mars.** *(JPL)*

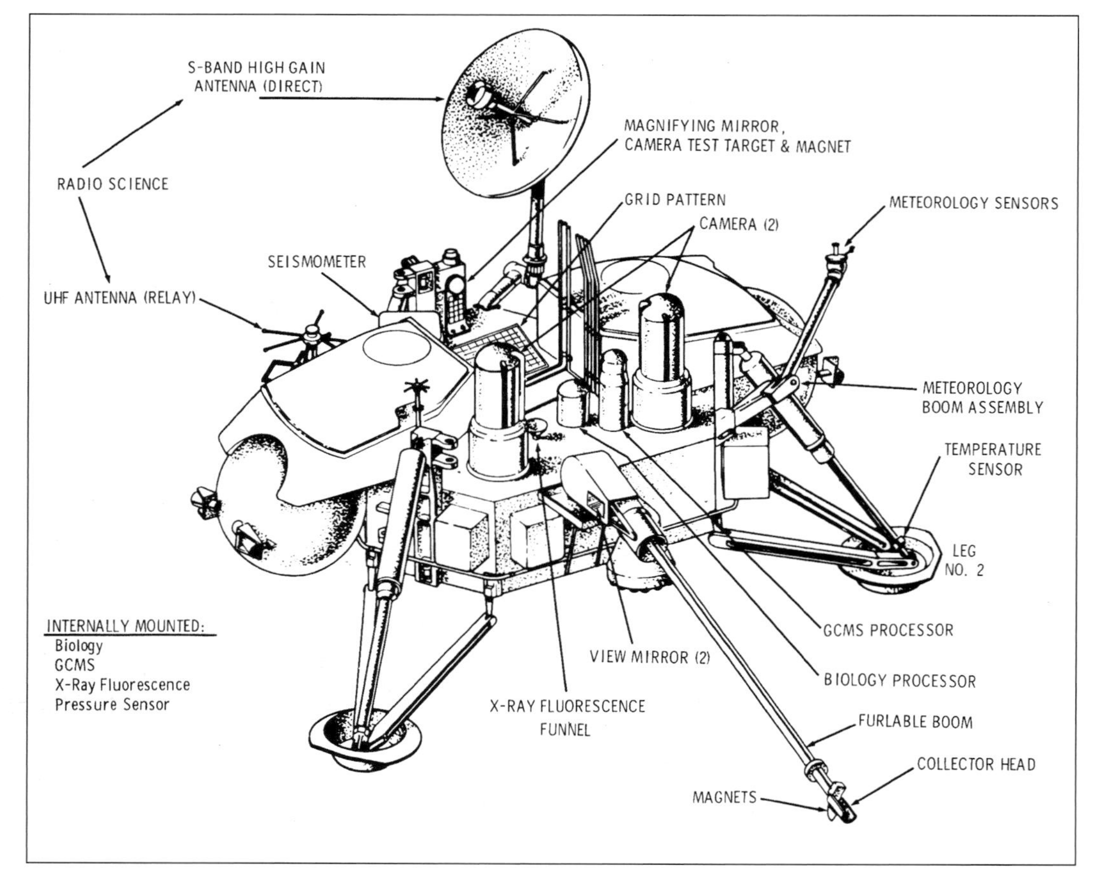

RIGHT **Lander science instruments included a biology experiment for testing samples using basic biochemical techniques for determining whether life exists in surface material.** *(David Baker)*

saw giant canyons greater in length than the continental breadth of North America, volcanoes three times the height of Mount Everest, and gullies and tributaries identical to those formed by fast-flowing water on Earth. There was evidence of glaciers having once crept down inclines, of mud flats caused by shallow rivers and of volcanic eruptions in the recent geologic past.

Next up was Viking, a mission involving two identical spacecraft each incorporating an orbiter and a lander, the latter encapsulated during flight by protective shells and riding piggyback on the orbiters. The plan was to place each spacecraft in Mars orbit and use powerful cameras to seek out definitive landing sites. Only then would the landers be released for descent to the surface, each one assigned a carefully selected site. While the general area chosen as the desired landing site for each spacecraft was a judgement based on previous Mariner pictures, much better images from the Viking orbiter cameras would narrow the specific landing spot to one where terrain was relatively flat and free of large boulders.

Launched in August and September 1975, Vikings 1 and 2 landed on Mars in July and September 1976, using a technology far more advanced than that applied to Mariner 4 a decade earlier. The two spacecraft went into orbit as planned and sent their landers to the surface in an attempt to search for signs of life using biological equipment sampling materials delivered by an extendible arm and scoop assembly. They sat on Mars for several years measuring their surroundings and recording weather conditions.

By 1983 NASA had selected a mission designed to gather information about the chemistry of the surface of Mars and the structure of its atmosphere. Initially known by the unwieldy title of Mars Geoscience/Climatology Orbiter, in 1985 its name was changed to Mars Observer, first in a series of missions NASA wanted to buy using 'off-the-shelf' (COTS) hardware, and industry was asked to submit bids to build the spacecraft. In October 1986 the then RCA Astro-Electronics company was awarded a contract to build Mars Observer, a hybrid of systems and components from two existing programmes, the Satcom-K

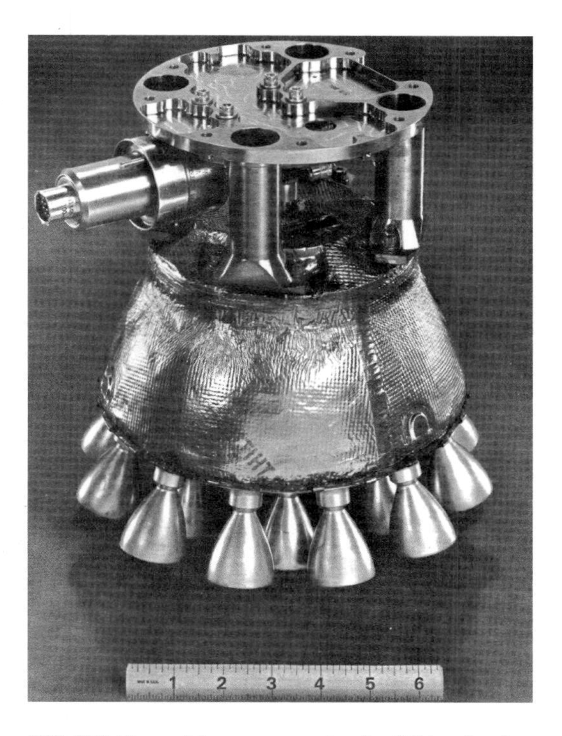

LEFT Each of three throttleable lander engines had 18 exhaust nozzles providing 62–638lb thrust to decelerate the spacecraft from about 135mph to a touchdown speed of 5mph. *(Rocket Research Corporation)*

BELOW The guidance computer for Viking landers was a complex piece of equipment programmed with instructions for autonomously guiding the spacecraft to a safe landing. *(Hamilton Standard)*

ABOVE **RCA engineer Ken Schmidt adjusts the lander's high-gain antenna employed to transmit images and data to Earth.** *(RCA)*

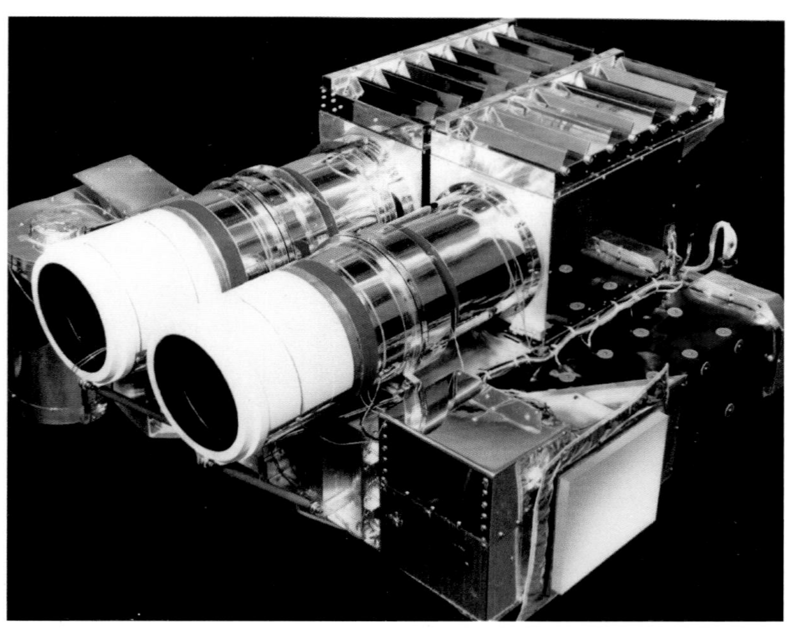

ABOVE **Attached to the scan platform on each orbiter, telescopic cameras were vital for providing pre-landing site verification images to refine the final spot selected for a landing, achieved on 20 August 1976 for lander 1 and 7 September 1975 for lander 2.** *(Ball)*

BELOW **The Viking orbiter carried a wide range of science instruments for sustained scanning of the surface while the lander was on Mars.** *(David Baker)*

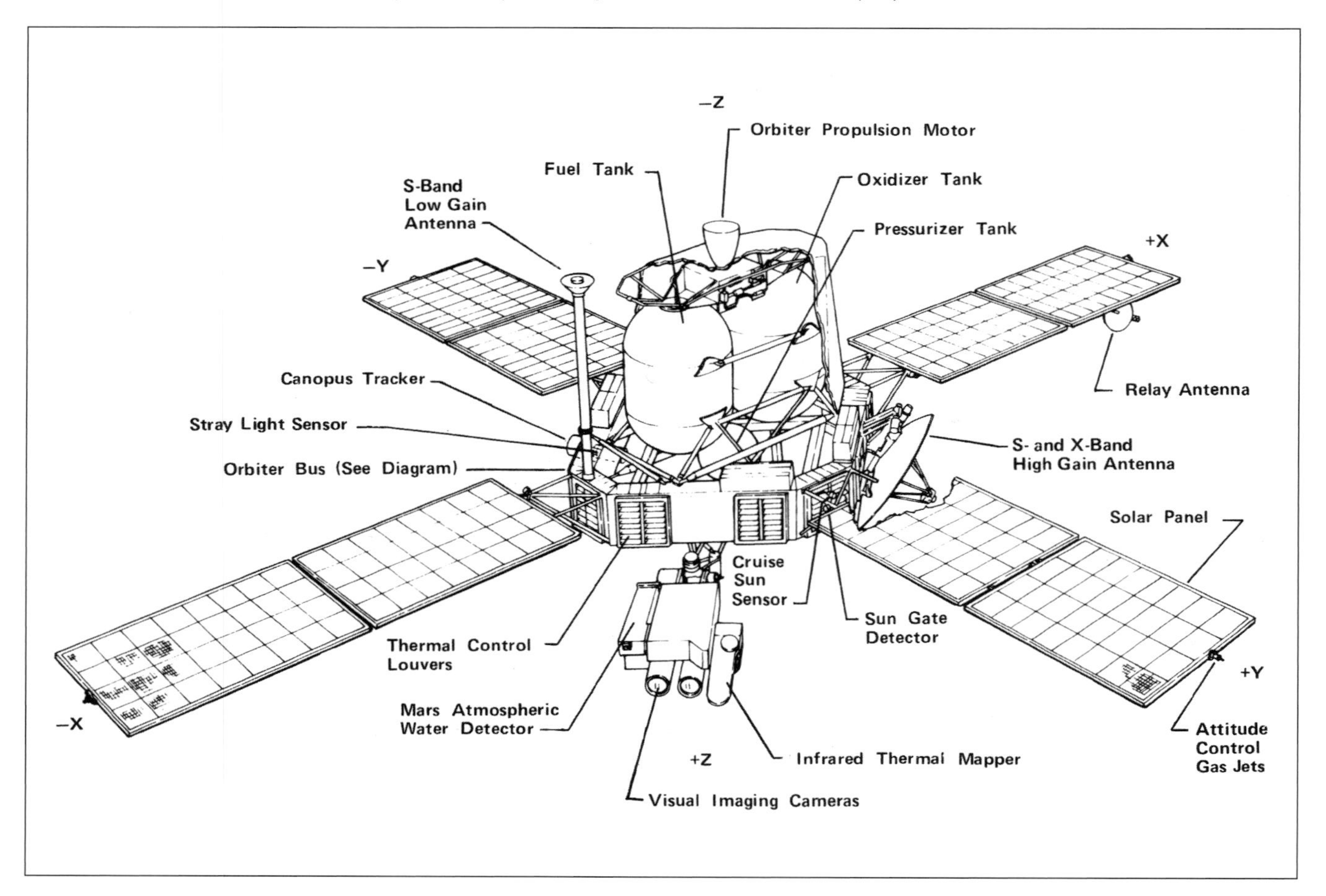

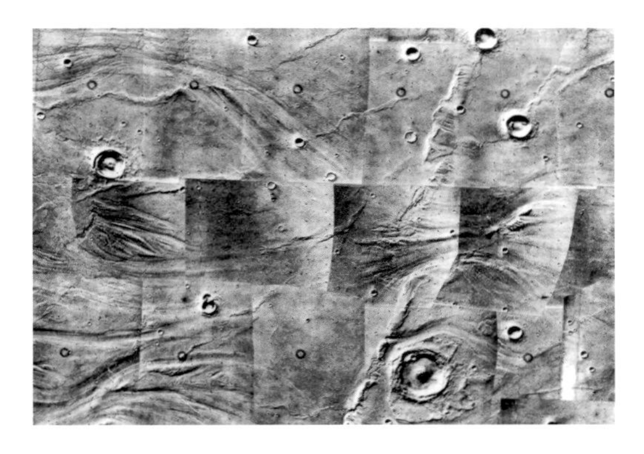

ABOVE **Mapped from orbit, a series of mosaic pictures assembles a meaningful interpretation of the terrain in Chryse Planitia, selected as the first landing site.** *(JPL)*

ABOVE RIGHT **The descent included atmospheric braking, a parachute and the powered descent phase.** *(Martin Marietta)*

BELOW **Viking's 53ft-diameter Dacron polyester parachute slowed the lander from 680mph to around 135mph.** *(Goodyear)*

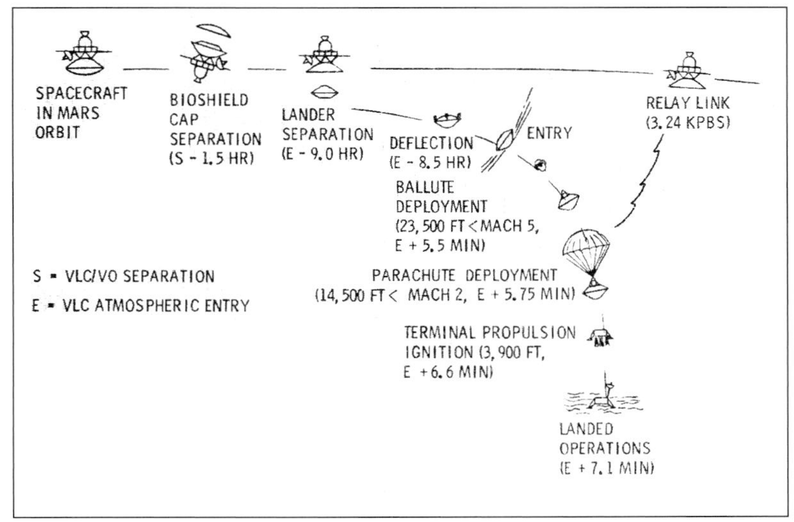

communications satellite and a family of weather satellites.

Managed by JPL, Mars Observer was to operate from a polar orbit and study the planet for an entire Martian year (687 Earth days) with a battery of seven instruments for thoroughly mapping the planet. But Mars Observer never reached its destination in working condition, failing before it arrived. Recovering from the disaster, NASA decided to fund a series of smaller low-cost missions launched at frequent intervals rather than big flagship missions once every decade or so.

The concept of a mini-lander put down on the surface of Mars carrying a small robotic roving vehicle gained acceptance, and a

BELOW **Viking on the surface! The first soft landing on another planet with a working spacecraft was hailed as a national achievement as America celebrated its 200th anniversary. Here, lander 2 surveys its surroundings.** *(JPL)*

radical plan emerged for putting a small roving vehicle down on the surface. Instead of a heavy mother-craft placing the entire assembly in orbit first and then releasing a lander to fly down to the surface on retro-rockets, why not use a direct flight straight from Earth into the atmosphere using a combination of atmospheric braking, parachutes, a small deceleration rocket and a set of inflatable airbags to protect the delicate rover as it bounced across the surface to a stop? Appropriately, it was called Pathfinder.

RIGHT A resident of Grovers Mill, New Jersey, Lew Chamberlain holds a model of Viking as he points to the location of the imagined invasion by Martians in Orson Welles' historic 1938 radio broadcast that stunned Americans into believing it was a real event! *(RCA)*

RIGHT The Viking lander 1 site bears evidence of trenching and sample retrieval for material tested in the biology instrument and a gas chromatograph mass spectrometer. *(JPL)*

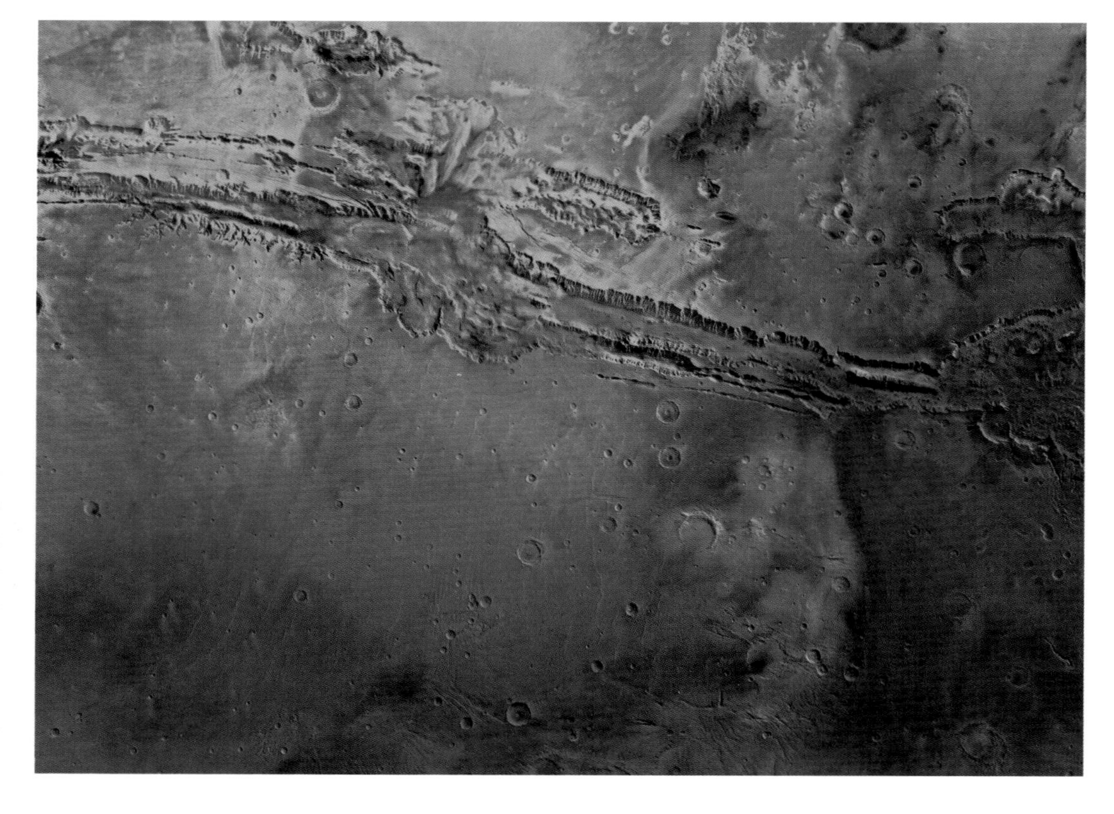

RIGHT The Viking orbiters provided the first really close look at Mars, including this enormous canyon (Valles Marineris) extending a distance equal to the continental spread of the United States. *(JPL)*

ABOVE The success of the Viking programme led NASA's Marshall Space Flight Center to capitalise on its management role with the Apollo Lunar Roving Vehicle to propose a tracked roving Mk 2 version. *(MSFC)*

LEFT Viking lander prime contractor Martin Marietta came up with a Viking Mk 3 that it proposed for launch in 1981, more capable and with greater mobility than the NASA version looked at during 1976. *(Martin Marietta)*

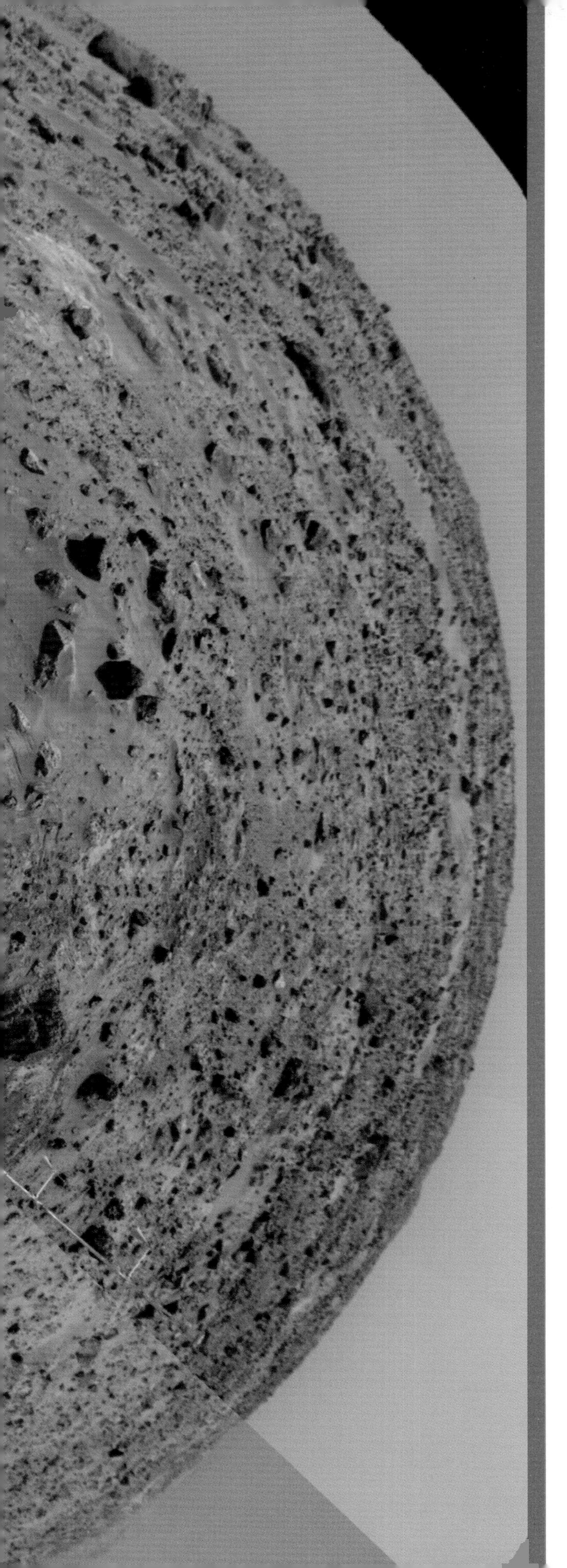

Chapter Two

A new generation

After early fly-by missions and two landers, the next step was to put a rover on the surface, a wheeled vehicle unlike anything built before. To survive in the hostile environment it needed new technology and a new means of operation, far from Earth and without human controls. The challenges were enormous but Pathfinder would live up to its name.

OPPOSITE A self-portrait by Pathfinder looking down from the Imager for Mars Pathfinder camera on top of the telescopic mast, showing the Mars Pathfinder lander, solar-cell petals and the ramp down which Sojourner travelled to the surface of Chryse Planitia. *(JPL)*

When Pathfinder was approved in 1993 nobody knew it would pioneer a way of putting rovers on the surface of Mars that would remain the benchmark concept for increasingly sophisticated vehicles over the next 20 years. Mars Pathfinder was indeed exploration on a shoestring but, as a true pathfinder for future and more ambitious missions, it was a flagship pioneer – and if it worked it would put the first working rover on Mars.

NASA decided on a direct flight straight to the surface, the cheapest and most energy-efficient means of getting there. Instead of first going into orbit like Viking, it would be sent on a trajectory that would align the flight path with the atmosphere so that it would encounter it at an angle of about 14°. Using a combination of atmospheric braking, parachute, rocket power and airbags, it would slice into the carbon dioxide atmosphere and slow down to an acceptable speed for a semi-soft landing – defined as somewhere between a destructive crash and a survivable impact. But it would take airbags to do it.

The Pathfinder spacecraft would travel inside an aeroshell with a backshell supporting a parachute system, a combination similar to that employed for Viking. Inside Pathfinder, cocooned within the pyramid-shaped structure, was the rover, called Sojourner. Instead of a separate spacecraft to first place it in Mars orbit, Pathfinder would be attached to a drum-shaped cruise stage providing propulsion for trajectory correction manoeuvres, solar cells for electrical power, equipment for receiving and transmitting communications signals from and to the Earth, and thermal protection for ensuring survival against extremes of temperature in the vacuum of space between Earth and Mars.

The cruise stage measured 8.5ft (2.65m) in diameter with a height of 5ft (1.5m). With a requirement for 178W of electrical energy to power its systems during the interplanetary phase, the cruise stage had solar cells across a total area of 30ft^2 (2.8m^2) of the upper (aft) Sun-facing surface. The cruise stage provided attitude control of the spin-stabilised structure, and mid-course corrections. It had in addition two Sun sensors and an adapter for mounting the entire assembly on top of the upper stage of the Delta II 7925 launch vehicle.

When launched, with the cruise stage attached, the assembly weighed 1,973lb (895kg) with 207lb (94kg) of propellant in four spherical tanks for eight 4.45N hydrazine thrusters in two clusters. The thrusters served both attitude control and course correction functions. The 2rpm spin axis was maintained to within 3° by the onboard attitude control system, the cruise stage itself being attached to the backshell by explosive bolts.

With a diameter of 2.65m the biconic aeroshell had a 70° forebody with a 49° aftbody geometry and would enter the atmosphere of Mars at a flight path angle of –13.8°, compared to –17.6° for the Viking Lander. With an entry mass of 585kg, the structure had an atmospheric 'footprint' of 62.3kg/cm^2 with no pitch, compared to the Vikings, which had a –11.1° angle of attack to achieve a nominal lift over drag (L/D) of 0.18.

The Pathfinder aeroshell had zero L/D and was spin-stabilised at a rate of two revolutions per minute on entry into the atmosphere compared with three-axis stabilisation for Viking. Because of these dynamics, a peak heating rate of 100W/cm^2 would be experienced with a maximum deceleration of 20g. Total heat load would be 3,865J/m^2, compared to 1,100J/m^2 for the Viking heat shield.

The Pathfinder parachute was a direct descendant of the Viking Disk-Gap-Band (DGB) type, a name derived from the description of its configuration when deployed: a bulbous canopy on top, a small gap below, and a cylindrical band to which were attached the lines from the circumference of the band down to the aeroshell. Development of the Viking parachute had the advantage of expensive full-scale testing in a wind tunnel and much of the work for that qualified the system in 1972. Engineers wanted to retain the performance and physics of the parachute without tinkering too much with the inflation dynamics. The parachute deployment incurred a dynamic pressure of 600N/m^2.

In the Viking parachute, the lines were attached to a 'swivel' point 7.2ft (2.2m) above the backshell, but with Pathfinder there was a riser holding the converged lines 6.5–13.1ft (2–4m) above the shell. Pathfinder's parachute had a diameter of 26ft (7.9m) and a height of 81.4ft (24.8m) from the centrebody of the aeroshell to the bottom of the band. This was shorter than the 98ft (29.8m) for Viking's 53ft (16m) diameter parachute. The Viking parachute was over-designed to operate between Mach 1.4 and 2.1 and a dynamic pressure of 400–700Pa. In reality this was a considerable margin compared to its operating value of Mach 1.1 and 350Pa, affording great flexibility for future DGB parachutes on Mars landers.

The descent velocity of the lander was reduced further by firing three Rocket Assisted Deceleration (RAD) motors, and for this a lander bridle system was essential for three vital functions: stabilising the dynamics of the suspended system during firing of the RADs; deploying the lander from the backshell; and releasing the lander from the backshell and bridle assembly. The two-body propulsion system did not have a control mechanism to maintain the thrust vector parallel to the lander's descent path, and thrust imbalances would tend to rotate the backshell around the lander's centre of mass.

Ideally, the lander would be separated from the RADs by the greatest length possible but line weight and stowage volume restricted that option. Based on tests and trade-offs between position and body motion, a bridle length of 66ft (20m) was considered optimum to stabilise the lander mass below.

Lander deployment from the backshell was effected using a Kevlar bridle with release rate controlled by a Descent Rate Limiter, or DRL, in effect the device used when abseiling from a rock face. The DRL used with Pathfinder was a slightly modified version of a product manufactured by Frost Engineering and had been designed as an emergency egress device for crew escaping from commercial aircraft. With changes to the lead brake weights, replaced with tungsten brakes, and removal of handgrips, the commercial off-the-shelf device was adopted in the best tradition of using low-cost equipment procured from an existing manufacturer.

The 350kg lander would unreel the full 66ft (20m) in ten seconds with the lander

BELOW Mars Pathfinder was a Discovery-class mission with the Sojourner roving vehicle attached as an 'experiment', to become the star of the show when Pathfinder landed on Mars on 4 July 1997. *(JPL)*

RIGHT The airbag concept for protecting the spacecraft originated in the US Army, the system adopted for Mars Pathfinder involving 16 bags, four for each of the four surfaces of the folded lander. *(JPL)*

descending at 164–246ft/sec (50–75m/sec). Release of the lander from the bridle was accomplished using a standard pyrotechnically actuated cutter produced by Roberts Research Laboratory.

The impact bag concept arose from US Army payload delivery systems and comprised a retro-rocket to brake the descent rate to a maximum vertical velocity of 22m/sec and a horizontal component of 72ft/sec (22m/sec) or 49mph and remain capable of surviving 10–30 bounces on rocks as high as 1.6ft (0.5m) on slopes of up to 30°. Impact bag inflation was triggered by solid propellant gas inflators fired two seconds prior to the firing of the RAD motors. The time of ignition and time of bridle release were computed so that the descent rate would be zero when the bridle was disconnected, ideally about 33ft (10m) above the surface. Impact with the surface was expected to be around 33–98ft/sec (10–30m/sec).

The 16 airbags were each constructed from two separate layers of polyester fabric, with each bag assembly containing a pressure vessel inflated by gas, and an outer layer covering about 60% of the surface for abrasion protection. When fully inflated the internal pressure of the bags was approximately 2psi, versus the 0.1psi of the atmosphere on Mars. Each gas generator had two different propellant grains, the first bringing the bags to full inflation rapidly, the second slower-burning charge sustaining pressure for up to two minutes, sufficient time for Pathfinder to bounce and roll across the surface.

Because the communication antenna was inside the closed lander and could not send signals to Earth, Pathfinder had a closed-loop fully autonomous system for deflating and retracting the landing bags and deploying the petals. The spacecraft could come to rest in one of five positions: base down, side down (any one of the three) or nose down. It might have rolled to a stop lodged against a rock, or it might be free to establish its centre of mass, base side down to the surface. Each of the four petals (three sides and the base) was protected by four large six-lobe airbags made from Vectran fabric, similar to Kevlar. Obviously, the lander would be resting on one set of deflated bags, and Airbag Retention Actuators (ARAs) would retract the other three.

Retraction cables acting like drawstrings were attached inside the bags to simultaneously open vent patches and cinch the bags on command. Five retraction cables, each about 18ft 8in (5.5m) in length, retracted into a spool inside the ARA. After retraction the gravity vector was sensed by the lander and, if it was laying on one of the three side petals, it would begin opening that petal to flip it over base down. The least likely orientation is nose down but in that eventuality, because it is highly stable in that position, it is necessary to shift the centre of mass over the pivot point to flip it over, and for that the airbags are used. The airbag against which the lander is leaning is left unretracted and the other bags are drawn in close around the lander. The two petals opposite the unretracted bag are opened only 20°, raising them to a vertical position 90° to the base for maximum height. The final bag is then retracted and the drag effect would pull down the lander on to its side, from which a side-recovery procedure would be started by the computer.

The ARA posed several problems for designers because of its operating environment. The device had to be a high-torque winching mechanism, be compact, and weigh as little as possible. It consisted of two DC brush Maxon motors each driving a 1,550:1 five-stage planetary gearbox connected to a single cluster spur gear, driving a secondary spur and spool assembly with an end-to-end gear ratio of 8,277:1. Each rotor had a stall, torque close to 15.8mNm (2.24oz-in) with each actuator given an optical encoder to 'count' motor revolutions that were an indicator of the amount of cable length retracted.

The same gearbox, with a customised ring gear and output, was also used in the gimbal device for the imager on Pathfinder and was selected for the robotic arm of Mars Polar Lander set for launch in 1998. In its application to the ARA, however, the fifth-stage planet carrier was a bearing-supported pinion gear driving the cluster gear. Final output stall torque of each assembly was 11.3Nm (100lb-in) and the total torque of the device was approximately 96Nm (850lb-in) at operating temperature.

Getting the three triangular-shaped petals open on the lander base was the responsibility of the Lander Petal Actuator, a large high-torque mechanism used to flip over and open the 350kg lander. It consisted of an ultra-high torque (12,000lb-in) device in a small package with minimal mass. Electronic commutated, brushless DC motors were selected for their high torque and great reliability, but also because they were available for another project and so were cheaper and accessible. The motor had a 0.14Nm (20oz-in) stall torque

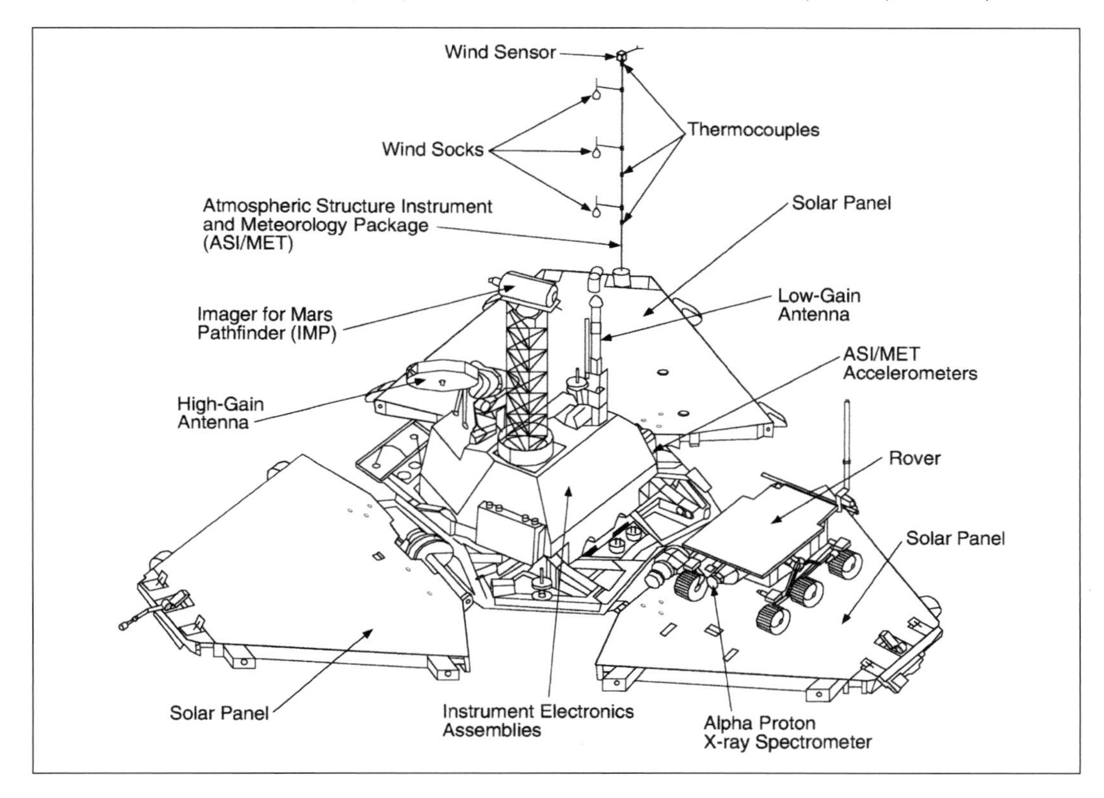

LEFT The three petals that encased the lander components on a central platform upon which it and the rover would come to rest after airbag deflation, the Sojourner rover being attached to the inner face of one petal. *(David Baker)*

driving a three-stage 49.3:1 spur gear head, which in turn drove a 4:1 internal spur gear set inside the actuator itself. This put power through a detent clutch to a 160:1 harmonic drive limited by a ratchet torque to 1,580Nm (4,000lb-in). The torque was transmitted to one hinge per petal via a titanium tube with square drive holes in each end. Output speed of the 31,552:1 gear ration actuator was about 40 minutes per revolution.

In all, there were 53 distinct pyrotechnically operated devices, including separation nuts, cable cutters, pyrovalves, pin pullers and gas generators, each powered by a redundant NASA Standard Initiator (SI). The 1/8in cable cutter is a standard device operated by pressure generated from a SI which drives a sharp steel blade through a material such as a cable or a cord, severing it as it strikes an anvil on the opposite side. Separation nuts, of which there were 21 on Pathfinder, are piston-activated devices using a segmented and threaded collet to hold interfacing bolts. They were used for separating the cruise stage, releasing the heat shield, releasing the lander, and the petal latches.

The lander was powered by three solar-cell arrays arranged at 120° spacing, with a total area of 30ft^2 (2.8m^2), giving up to 1.2kW hrs/day on clear days, a supply provided at night by rechargeable silver-zinc batteries providing more than 40 amp-hours at the beginning of the surface mission. The lander's flight computer was a slightly modified commercial R6000 32-bit microprocessor operating at a speed of 20MHz with 128MB of dynamic RAM and 4MB of non-volatile memory. Using the C-program language, flight software controlled all spacecraft functions and command and telemetry processing. Communication with the Earth was made using the radio-frequency telecommunications subsystem with RF and antenna subsystems. The system operated at X-band (7.2–8.5GHz) frequencies with a small 5W backup transmitter and telemetry modulation unit in the event of a malfunction with the primary transmitter.

During the cruise phase between Earth and Mars, the radio system used a medium-gain antenna, providing uplink and downlink telemetry rates of 250bps and 40bps, respectively. A low-gain antenna on the lander could support an 8bps uplink and up to a 300bps downlink. A steerable high-gain antenna could support 250bps uplink and up to 2,700bps on a downlink. The lander carried a single camera known as the Imager for Mars Pathfinder, or IMP. With a mass of 11.5lb (5.2kg) it consumed 2.6W of electrical power and was supported on top of a telescoping mast. It had two optical paths for stereo imaging, each carrying a filter wheel providing 12 colour bands in the 0.35–1.1 micron range, exposures through the various filters combining to produce colour pictures. The IMP's field of view was 14° in both horizontal and vertical axes and it was capable taking one frame of 256 x 256 pixels every two seconds. The mast had a deployed height of 1m.

The lander also had an Atmosphere Structure Instrument/Meteorology Package (ASI/MET) mounted on a separate arm. It gathered atmospheric data during descent of the lander through the atmosphere, acquiring information on temperature, pressure and wind levels down to the surface and for the duration of the mission. Temperature was measured by thin wire thermocouples mounted on the mast deployed after landing, spaced at intervals of 10in (25cm), 20in (50cm) and 100in (16cm) above the surface. Pressure was measured by a Tavis magnetic reluctance diaphragm not unlike a similar instrument on the Viking landers. The wind sensor employed six hot wire elements distributed uniformly around the top of the mast.

Designing Sojourner

Pathfinder's rover was called Sojourner, a name chosen in July 1995 by a panel of judges from the Jet Propulsion Laboratory and the Planetary Society following a worldwide competition. Students up to 18 years of age were invited to select heroines and submit essays explaining their historic accomplishments. Valerie Ambrose of Bridgeport, Connecticut, chose Sojourner Truth, an African-American liberationist and a campaigner for civil rights during the American Civil War of 1861–65. Shortened to Sojourner, the name also means 'traveller'.

Sojourner was mounted to the lander in a

kinematic arrangement of two bipeds in the vertical and lateral axes, an additional biped in the vertical axis and one shear pin in the lateral axis. The bipeds and monopod were preloaded to the body via stainless steel cables tensioned to 1,000lb (453kg) and 400lb (181kg) respectively. When the cables were cut the restraints were free to un-stow by driving the rear wheels forward while locking the others so that the body rose off the shear pin. This induced a rotation on one rocker-half relative to the other, disengaging a tab that allowed a spring-activated antenna to deploy. At the top of the travel a spring latch tied the two pieces each comprising a rocker-half forming a rigid suspension system. To completely disengage the wheel restraints, the rover backed up slightly and was free to drive.

The rocker-bogie suspension was unique in that it did not use springs, but this design would be used for all NASA Mars rovers and was perfectly adapted to the rough and undulating surface of Mars. With six wheels rather than four, its joints rotate and conform to the contour of the surface, providing greater stability and ease when crossing obstacles. Each wheel would be effectively positioned by its attendant bogie arm to assist a separate wheel overcome obstacles larger than itself. Vehicles with six wheels can surmount obstacles three times larger than those crossed by a four-wheel chassis. To this day, Earth-based off-road vehicles and SUVs have failed their owners in not seizing this logical form of mobility and are thus limited in what they can achieve.

The origin of the six-wheeled bogie lies with several people at JPL who since the early 1960s had worked on trying to fathom out how to build a mobility system for rough terrain, and, in addition, give it a level of autonomy so that it could be left safely to roam on its own far from contact with Earth – such as on Mars.

ABOVE Development of the micro-rover began in the late 1960s during research programmes which eventually led to Sojourner, only funded as an experimental tool of a technology demonstrator. *(JPL)*

LEFT The success of Sojourner would be largely due to the rocker-bogie system whereby six wheels would wrap around obstacles and distribute the frictional coefficients through the wheels where normally they would be absorbed by the chassis. In the rocker-bogie concept, the chassis hangs on the suspension. *(JPL)*

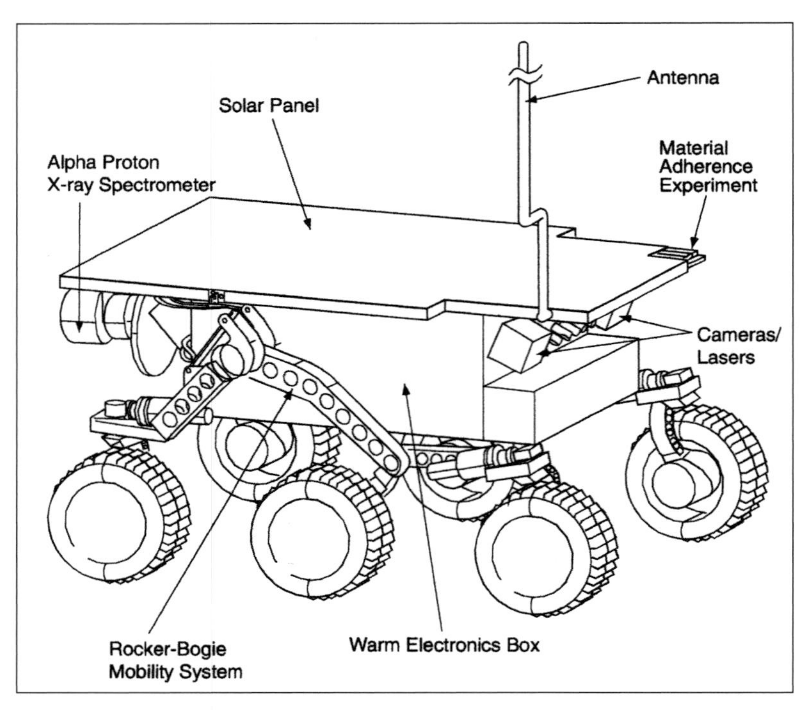

ABOVE The design emphasis on the 23lb Sojourner was to demonstrate several enabling technologies that could be applied to future, more capable and longer-lived rovers such as Spirit and Opportunity, launched in 2003. *(David Baker)*

But two people stand out for their brilliant engineering skills and ingenuity: Don Bickler and Howard Eisen. Bickler took an existing pool of research and experimentation on a wide variety of mobility concepts for rovers and decided they would not work on Mars. Long before Pathfinder was a project, Bickler came up with a six-wheel rocker-bogie concept that worked to perfection.

Sojourner was engineered to tip up to 45° without falling over. Each 5in (13cm) diameter aluminium wheel had a width of 3.1in (8cm), with a stainless steel tread and cleats for maximum traction, operating independently of the others. Three motion sensors would detect excessive tilt and inhibit motion, but Sojourner could scale a rock more than 8in (20cm) high and continue on its way. With a surface pressure on Mars of a mere 0.5lb/in^2, the wheels had a track of 15in (38cm) with a wheel base of 10in (25.4cm) between the front and middle wheels and 9in (22.8cm) between those and the rear wheels.

Because of strict quarantine restrictions on the transfer of Earth microbes to Mars, care had to be taken in the selection of materials from which the rover and its mobility was built. All electronics were sealed inside a Warm Electronics Box, or WEB, insulated and with a copper wire as a heat link. Three radioisotope heater units (RHUs) each contained about 2.6g (0.1oz) of plutonium 238 providing 1W of heat to keep the electronics box warm against Mars's frigid temperatures.

It was necessary to minimise the wiring interface, and a Maxon brush DC motor was chosen for its unique capacitor commutation, where a common capacitor between the segments allowed inductive energy stored in a reversing motor segment to dissipate in a circuit instead of an arc. In a Mars environment, where Paschen voltage is low and arcing is easy, this was an obvious choice and played a major part in reducing anti-commutator wear. The permanent magnet was a two-wire device without temperature-sensitive components.

JPL selected an LED/phototransistor pair with a rotating mask to create a one-bit relative encoder for odometer readings, the mask, detent rotor and pinion gear being attached directly to the motor output shaft. In magnetic detent and with power off the motors, the rover remained static through holding torque as though it were a parking brake. The gearbox selected was a sintered-metal planetary system manufactured by Globe Motors. JPL engineers reduced its weight and modified the gear and housing by shortening the last-stage pinion and sun gears and by placing miniature ball bearings between the planetary gears and the shafts of the first two high-speed stages. The gears were vacuum impregnated with low viscosity oil and, because the drive ratio was 2,000:1, each encoder corresponded to one revolution of the motor for 0.008in (0.2mm) of wheel motion.

The steering actuators were identical in drive but the gearbox had a 2:1 right-angle bevel gear to accommodate engineering requirements for mounting the rover inside the lander and getting it to fit. The steering position was read via a conductive plastic Beckman potentiometer on the bevel gear. The assembled drive actuator weighed a mere 180g and had an offloaded speed of 1.5rpm and a stall torque of 70lb-in at 15.5V. Very low temperature actuator testing demonstrated that as temperatures decreased the motor resistance dropped and stall torque increased to 160lb-in, and a succession of long-duration life tests indicated that the motors could accommodate unusual and varied loads and exceed 10km in driving distance. As

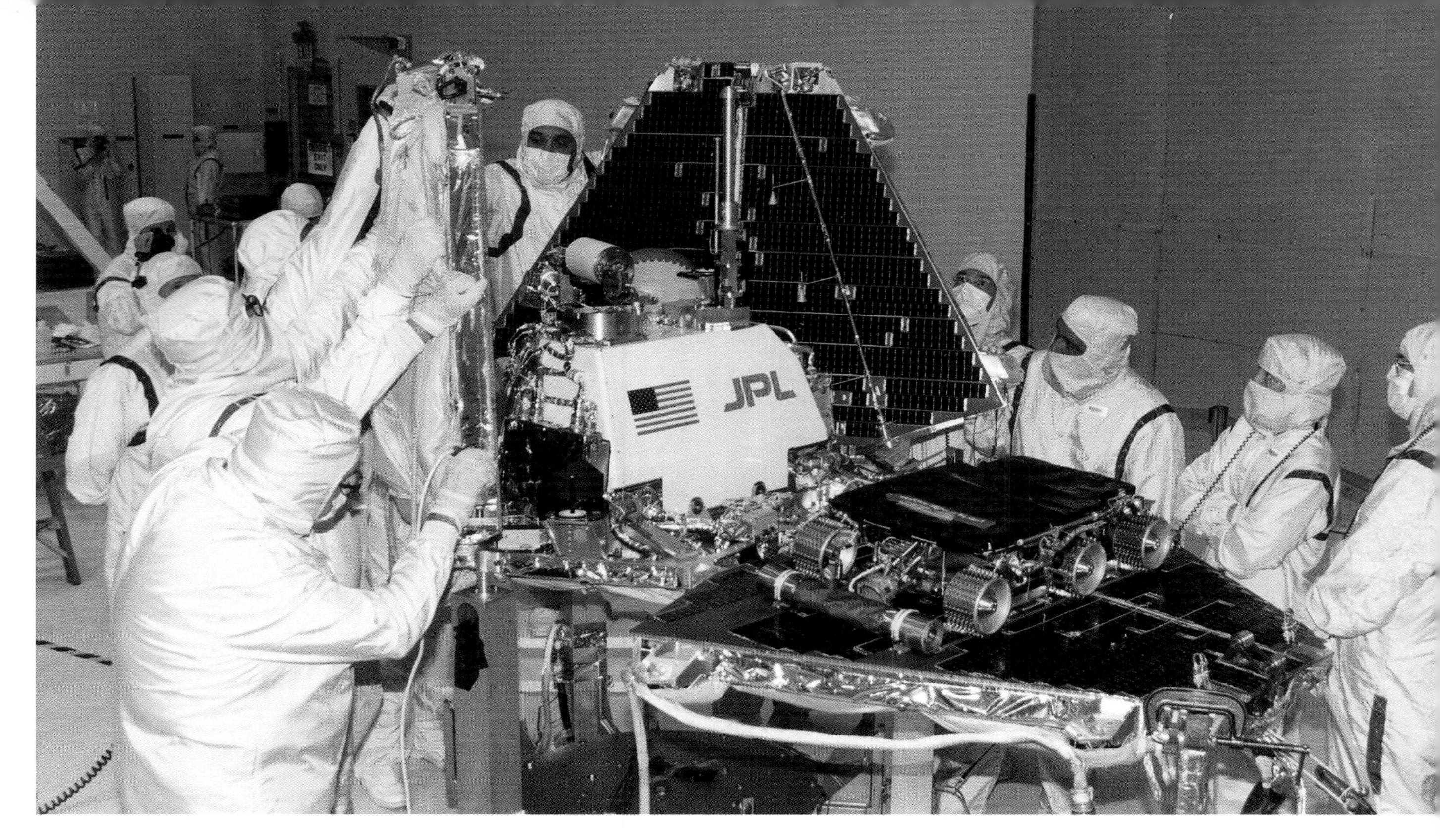

insurance, heaters were installed on the motors to decrease the draw on the current during cold early morning hours but they were never actually used during surface operations at Mars.

To get off its mounting plate on one of the Pathfinder petals to the surface, Sojourner needed a ramp, for which there was a rigid specification. The ramp had to be folded for flight and contained within a very small space until needed. It had to deploy at inclinations of up to 30°, support a rover weighing up to 27.55lb (12.5kg), resist buckling in a cantilevered position when the rover was between $1/3$ and $2/3$ of its length, and extend without obscuring or shadowing any of the solar cells on the three lander petals. To conduct a full design, fabrication and test programme on a deployment mechanism, JPL contracted with Astro Aerospace Inc. to come up with a workable and highly reliable system based on the company's experience with deployable communication masts.

The chosen method was a deployable assembly rolled up into a compact cylinder 16.5in (42cm) long and 3in (7.62cm) in diameter. It was only by this means that Sojourner could drive down on to the surface; so to insure against one direction being blocked by ruffled airbags or rocks, two would be provided – one directly in front of Sojourner and the other behind. Preloaded, and released with cable cutters, the ramps would unfurl to a length of 53.5in (136cm) and, with a width of 16.5in (41.9cm), they provided an unimpeded route to the surface – in effect a self-rigidising structure. Depending upon surface obstructions front or back, controllers could select Sojourner to drive forwards or backwards off the lander. To prevent Sojourner falling off sideways a small inset curb each side of the ramp helped guide the rover's wheels, and flat stainless steel sheets were provided for the rover to drive upon. Each ramp weighed less than 2.2lb (1kg).

When deployed on the surface of Mars, Sojourner was 11in (28cm) high, 24.8in (63cm) long and 19in (48cm) wide, with a ground clearance of 5in (13cm), its box structure being mounted on its rocker-bogie suspension system. On the Pathfinder lander it had a height of 7.1in (18cm) until it was released from its compressed position, essential to fitting it inside the limited volume within the tetrahedron-shaped folds of the petals. At launch Sojourner weighed 34.2lb (15.5kg), but, once deployed from Pathfinder, on the surface of Mars it weighed 23lb (10.6kg).

The onboard control system was an Intel 80C85 8-bit dual-speed central processor with a capacity to run about 100,000 instructions per second, its computer capable of compressing

ABOVE The size of a small microwave, Sojourner was squished down for attachment to the lander, its mobility system collapsed down to reduce overall height, allowing it to fit within the pyramidal-shaped interior of the enclosed Pathfinder spacecraft. Note the rolled ramp retainer and the thin, flattened, airbags on the vertical petal at left of picture. *(JPL)*

OPPOSITE Launch-pad technicians prepare to close the payload shroud encasing the terminal stage of the launch vehicle, the Pathfinder cruise stage and the conical aeroshell, which unlike Viking did not have a bioshield. *(NASA-KSC)*

and storing a single image. Sojourner had 512 Kbytes of RAM and a transmission rate of 8Kbps. Communication with the lander was effected through a Motorola link with a range of 1.2 miles at 9,600 baud with a 39.4in (1m) whip antenna.

The single 1.9ft^2 (0.17m^2) array of solar cells on a panel across the top of the rover provided power for up to seven hours a day, with backup provided by a non-rechargeable lithium thionyl chloride D-cell battery. The gallium arsenide array was capable of providing 16W of power at midday but only for a span of around four hours, centred on noon. At other times power was a meagre 11W. In fact, there was insufficient power to simultaneously operate the radio frequency modem and take images, to run all four steering motors and the six drive motors, or to run the six drive motors and operate the camera/laser-based obstacle detection system at the same time.

All communications would go through Pathfinder because Sojourner had what was little more than a walkie-talkie to send and receive messages and data via the lander. Because the emphasis was on low cost, Sojourner's radio was a commercial off-the-shelf product of the kind that could be purchased at a local radio store. Essentially a modem, the radio transmitted short bursts of data symbols, each 'packet' containing 2,000 eight-bit bytes. It was through these packets that rover images and data would be sent across to the adjacent lander for retransmission to Earth, using the lander as a relay.

The roving vehicle's radio consisted of two parts, a digital portion on a printed wiring board and an analogue portion on a separate circuit board. The digital component acted as an interface between the analogue board and the computer inside Sojourner, processing data for transmission or reception. To transmit data the analogue board would switch on its UHF transmitter and send modulated radio waves corresponding to the digital information formatted by that board. To receive commands via the lander, the analogue board would tune to the same frequency, amplify and filter them and demodulate the digital signals so that the digital board could input each information bit within a packet as it was received.

The Sojourner modem had a mass of 106g and was 3.2in (8.13cm) x 2.5in (6.35cm) x 0.9in (2.3cm) in size, operating on a radio frequency of 459.7MHz with a bandwidth of 25KHz and a transmitter power of 100mW. It operated on a 7.5V current. Maximum data rate was 9,600 bits/sec with an effective rate of 2,400 bits/sec. Sojourner's monopole antenna had a length of 17.7in (45cm) and stood up when the rover was unlatched from its compact position on the lander petal. With Sojourner off the lander, the antenna assumed a height of 32.7in (83cm) above the surface.

The equivalent radio carried aboard the Pathfinder lander was essentially similar to that installed in Sojourner, with the exception that it incorporated an extra electronics board, as the original was meant to be powered by a 28V source. It also had internal heaters to maintain battery temperature during the cold Martian nights. It had two DC connectors, one for power and one for signals. The radio was attached to the lander battery case and had a weight of 265g, with a size of 4.2in (10.6cm) x 2.8in (7.1cm) x 2.1in (5.3cm). It operated on a 28V DC current at 1.5W and had the same 100mW transmitter power as the modem installed aboard Sojourner. The fibreglass Pathfinder antenna had an overall length of 13.2in (33.6cm) and the system had a bandwidth of 16MHz at the 459.7MHz frequency.

The first rover on Mars

The 1996 launch window to Mars was used for the launch of two NASA spacecraft: Mars Global Surveyor (MGS) and Mars Pathfinder. First away, MGS was launched on 7 November 1996 and propelled to a Type II trajectory that would take it 309 days to reach Mars, at which time it would fire an onboard rocket motor to place itself in a highly elliptical orbit from where it would gradually be brought closer to the planet for an extended mapping operation. Using a special camera designed and developed by Malin Space Science Systems, it incorporated three instruments, the most powerful of which was capable of imaging surface features with a resolution of 4.9ft to 39ft (1.5m to 12m) per pixel. MGS also carried a laser altimeter, a thermal emission spectrometer and a magnetometer. Built by Lockheed Martin, MGS weighed 2,272lb (1,030kg) at launch.

Pathfinder/Sojourner was launched on 4 December 1996, encapsulated in the shroud of its Delta 7925 launch vehicle. The third stage placed the spacecraft on a near-perfect Type I trajectory that would take it 212 days to reach Mars, and the mission settled down to a long flight through interplanetary space. The terminal rocket stage that placed Pathfinder on course had spun the stage and its payload to 12.3rpm, and on 11 December the cruise stage – now separated from the rocket stage – reduced that to 2rpm, which it would maintain all the way to Mars.

Four separate trajectory correction (TCM) manoeuvres were made. On the first TCM, by far the largest, conducted on 9 January 1997, two of the spacecraft's eight thrusters fired for 90 minutes, changing velocity by 102ft/sec (31m/sec), or 69.5mph. The second TCM was performed on 3 February as a two-part manoeuvre developed by JPL engineer Guy Beutelschies, the plan being to move the flight path closer to Mars. In the first part the spacecraft fired its two forward-facing thrusters for five minutes, changing velocity by about 4.9ft/sec (1.5m/sec), or a mere 3.3mph. In the second part that day all four thrusters fired for five seconds, altering speed by 0.33ft/sec (0.1m/sec), or 0.2mph, and fine-tuning the path.

Being on a longer and slower path to its destination, Mars Global Surveyor was overtaken by Pathfinder on 14 March, 27 million miles (43.7 million km) from Earth and 43.2 million miles (69.7 million km) from Mars. The third Pathfinder TCM took place on 6 May when three burns tweaked the trajectory, leaving a few errors that were erased on TCM-4 conducted on 25 June just nine days before arriving at Mars. In a two-part event, the first burn lasted 1.6 seconds and involved four thrusters; the second burn 45 minutes later fired two thrusters for 2.2 seconds. The net effect was to micro-tweak the velocity by a fractional 0.06ft/sec (0.018m/sec) or 0.04mph, demonstrating the ultra-accuracy demanded by a direct entry into the atmosphere at just the right angle. The opportunity to carry out a final trajectory correction manoeuvre could have

been performed either 12 or 6 hours prior to entry but was felt unnecessary.

About 45 days prior to entry ground controllers shifted into high gear, preparing the spacecraft for entry and landing. A final health status review was undertaken, when controllers checked out Pathfinder's instruments and the Sojourner rover on 19 June. It was then that Pathfinder was activated operationally – until then it had been in dormant mode – with some software changes updating the processor. On 23 June flight controllers started loading 370 command sequences so that Pathfinder would have the latest updated information to autonomously control its own entry, descent and landing (EDL) events. The 124 x 62-mile (200 x 100km) landing ellipse was centred on an area known as Ares Vallis, 19.3°N by 33.6°W, and the combined cruise stage and spacecraft were heading right down the middle, slicing toward the atmosphere at a near ideal approach path of –14.06°.

Events prior to atmospheric entry began with venting the heat rejection system used to cool the spacecraft during cruise. About 30 minutes prior to entry the cruise stage separated, and the conical aeroshell and heat shield were on their own. Entry occurred five minutes prior to landing (L–5), at a height of 81 miles (130km) and a speed of 24,510ft/sec (7,470m/sec) or 16,700mph. Peak deceleration came two minutes later, exerting a 16g load on the decelerating spacecraft. Parachute deployment occurred at L–2min 14sec and a velocity of 1,214ft/sec (370m/sec), and the heat shield popped off 20 seconds later when its three separations fired, ejecting it at a velocity of 3.3ft/sec (1m/sec). Lander separation from the aeroshell came at L–1min 34sec as the spacecraft unwound to the full extent of the 66ft (10m) bridle.

The radar altimeter acquired the surface at L–28.7sec, at an altitude of 1 mile (1.6km) and descending at a velocity of 223ft/sec (68m/sec). The radar successfully held lock all the way through rocket ignition. The 16 airbags were inflated at L–10.1sec when the spacecraft was just 1,165ft (355m) above the surface. But Pathfinder's descent rate was somewhat higher than predicted and the RAD rockets were not ignited until the spacecraft was about 321ft (98m) above the ground, at L–6.1sec and a velocity of 200ft/sec (61.2m/sec). The bridle connecting the lander to the aeroshell was cut at L–3.8sec, just 70.5ft (21.5m) above the surface, and the spacecraft encased within its assembly of airbags floated down to the ground.

Initial impact occurred at 9:56.55am Pacific Daylight Time on 4 July 1997 at a vertical speed of 46ft/sec (14m/sec) or 31.3mph (50.5km/h), and an initial deceleration of 18.6g, resulting in the bagged spacecraft bouncing back up to a height of 50ft (15.2m) followed by at least a further 14 bounces over the next two minutes before coming to rest about 3,100ft (1km) away.

On Mars it was still dark, just 2:58am in the morning local time. The lander would operate on battery power until the petals had been deployed and the Sun got up. Touchdown on Mars had been achieved on the 221st anniversary of the foundation of the United States. The landing spot was a mere 16.8 miles (27km) from the bull's eye at the centre of the targeted ellipse, 1.55 miles (2.5km) below the mean surface radius of the planet so as to take advantage of a slightly denser atmosphere at deployment of the parachute.

The EDL phase officially ended 1hr 37min after initial impact when the petals had been fully deployed and the lander began its surface phase. On Mars, it was early in the morning, and the Sun was rising over the eastern horizon when the first telemetry was received on the ground 4hr 12min after landing, confirming that the lander and the rover were in good health.

Science at the surface

From what many had predicted was an unsurvivable sequence of high-risk events, a new way of reaching the surface of Mars had proved to be sound after all. Commands went back from Earth to deploy the high-gain antenna, search for the Sun, take the first pictures from the IMP and downlink all that back home. When Pathfinder landed signals took 10min 35sec to get back to Earth, and all times quoted here are the times at JPL when that information was received. Mars was 119 million miles away (191 million km) and Pathfinder had travelled 309 million miles (497 million km) on its arching path around the Sun, moving outward all the time until it encountered Mars.

OPPOSITE First of the replacement spacecraft doing the work of the failed Mars Observer of 1993, Mars Global Surveyor carried an advanced mapping camera capable of resolving objects down to one metre in size. MGS played a major role in helping plan future landing sites for more advanced roving vehicles. *(NASA)*

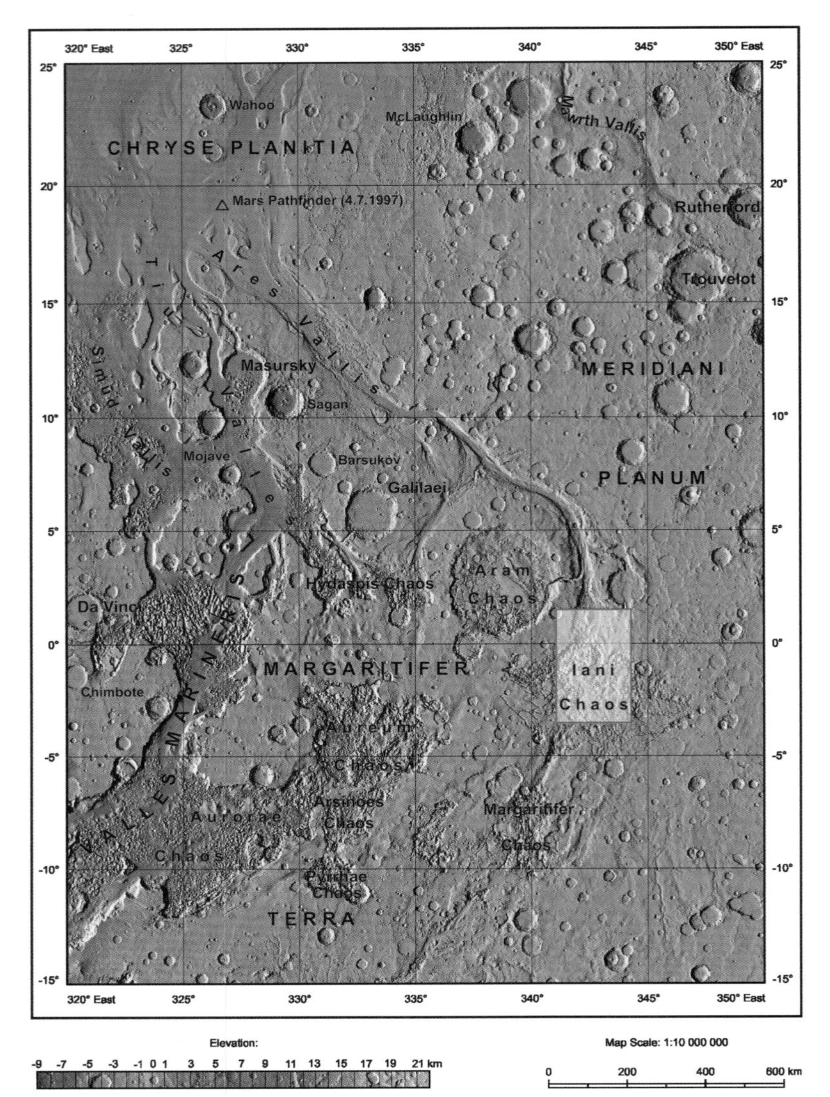

ABOVE The landing site for Pathfinder was the northernmost outflow of the Ares Vallis, where flow channels appear strong indicators of water-borne deposits freely flowing into the Chryse Planitia region and the northern lowlands. *(Malin Space Systems)*

Sojourner carried a special instrument to analyse surface soil and rocks, part of a cooperative arrangement with the Max Planck Institute for Chemistry at Mainz, Germany. This consisted of an Alpha Proton X-ray Spectrometer (APXS), a derivative of instruments flown on the failed Russian Mars landers of 1996. Attached to the Sojourner rover, the APXS device could take spectral measurements not only of the Martian dust but of rocks and outcrops to determine their elemental composition. Its electronics were housed in relative comfort inside the rover's Warm Electronics Box.

The spectrometer works by bombarding a rock with alpha particle radiation, charged particles equivalent to the nucleus of a helium atom consisting of two protons and two neutrons. The sources of the radiation are small pieces of the radioactive element curium-244 and, when fired at the rock surface, the alpha particles either interact with it and get reflected back or cause X-rays or protons to be generated. These 'backscatter' particles are counted at various energy levels, the number corresponding to the relative abundance of various elements in the rock, the energies measured being a product of the type of element encountered. It can take up to ten hours to get a high-quality result, the rover and its APXS remaining stationary and continuously bombarding the rock.

The first pictures from Sojourner revealed what scientists dubbed a 'rock garden' in front of the lander and two large hills on the horizon named the 'twin peaks'; of more immediate importance, it showed that the airbags under the rover petal had not fully retracted and would prevent it reaching the surface. Commands were sent to raise the rover petal 45° and resume bag retraction, which further images revealed was successful. The ramp was deployed, and although it did not reach the surface it was considered that the weight of Sojourner would bring it into contact with the ground. Sojourner's primary mission was planned to last seven days but Pathfinder was expected to survive for 30 days – more if everything went well.

Some 24hr 37min long, days on Mars are referred to as sols, Sol 1 being the day Pathfinder landed. On Sol 2 Sojourner was commanded to drive down the ramp, which it accomplished at 8:30pm PDT on Earth, 34hr 33min after landing, taking sample images to check out the camera system on the rover and the ability of the rover to communicate with Earth via the lander. On Sol 3 Sojourner moved further away, obtaining an APXS measurement of a rock dubbed Barnacle Bill, following that on the next several days with a rock called Yogi, acquired on Sol 9. By this time the prime objectives of the rover had been accomplished.

Devised as a technology demonstrator, it had done that – and more – but it was not done yet! More APXS measurements were made of rocks named Wedge, Shark, Half Dome, Moe and Stimpy, as Sojourner more than proved its worth as a mini-geology robot on the surface of Mars. Never designed to last long, after Sol 55

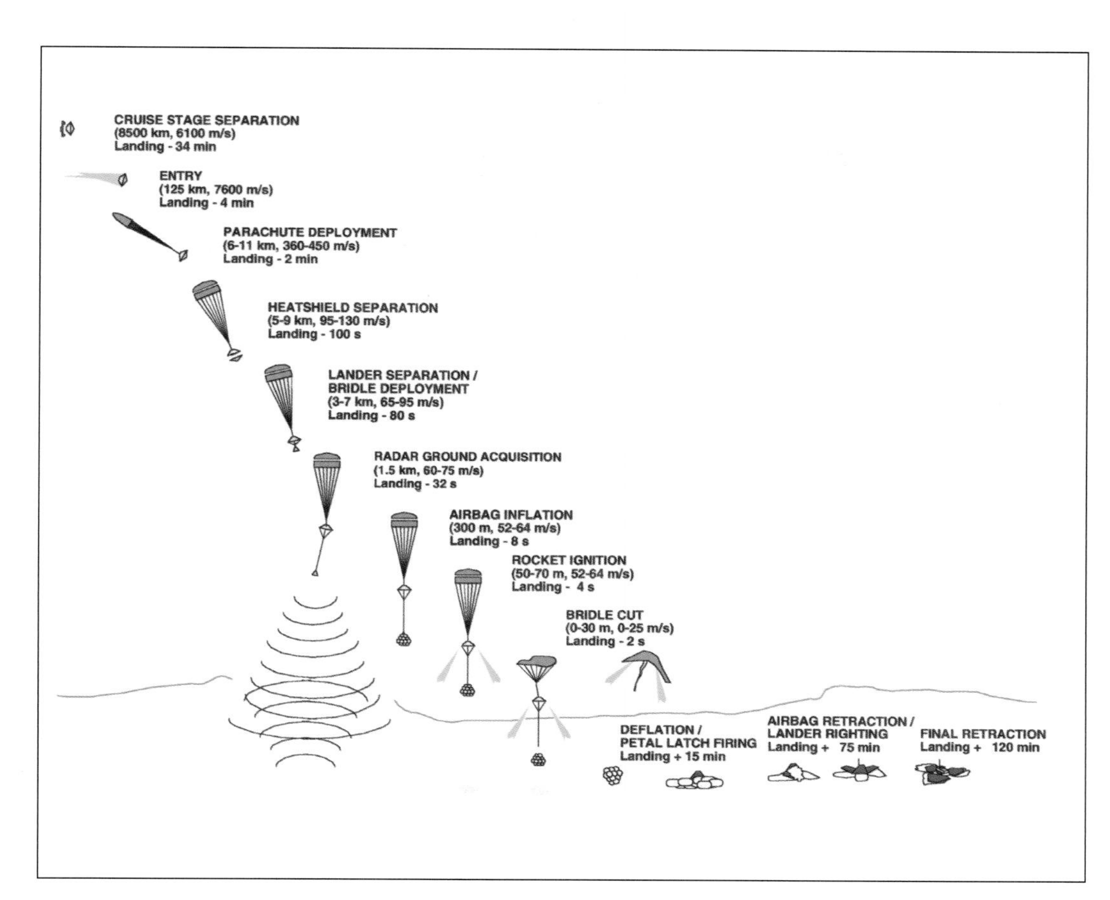

LEFT Following a direct entry into the atmosphere, Pathfinder took a decelerating ride using the atmosphere to exchange energy for heat before slowing for a supersonic parachute deployment and a rocket-retarded fall to the surface. *(JPL)*

only daylight activity could be conducted, as the batteries were by then fully depleted. By Sol 82 the days of Pathfinder/Sojourner were over. The batteries on the lander finally expired, and being unable to power the heaters essential to keep the electronics warm during the cold Martian nights they too failed. Without them communications could not be relayed to Sojourner.

But the small isotope heater pellets in the little rover kept Sojourner warm at nights, and during daytime hours energy from the Sun would power it up again. In the absence of hearing from Earth it would move back close to the lander and continue to circle it taking pictures and trying to send them via the lander, now silent and unable to talk to Earth or to receive signals from Sojourner. Only when completely covered with dust and no longer able to provide power from its solar panel would Sojourner come to rest, devoid of energy.

Initial science results were encouraging, with the APXS revealing more than one rock type at the site, an outflow channel of a catastrophic flood at a distant time in the history of the planet. Measurements of Barnacle Bill revealed a relatively high silicon content that could have been caused by weathering or by a form of differentiation not dissimilar to that which the Earth had been through early in its evolution. Results from Pathfinder confirmed a close affinity with expectations, based on extensive prior study of images from the Viking orbiters. Although relatively short-lived, Pathfinder encouraged further study with the remote

BELOW An engineering mock-up of the lander and a simulated Mars surface provided test data for refined design detail of the final concept. *(JPL)*

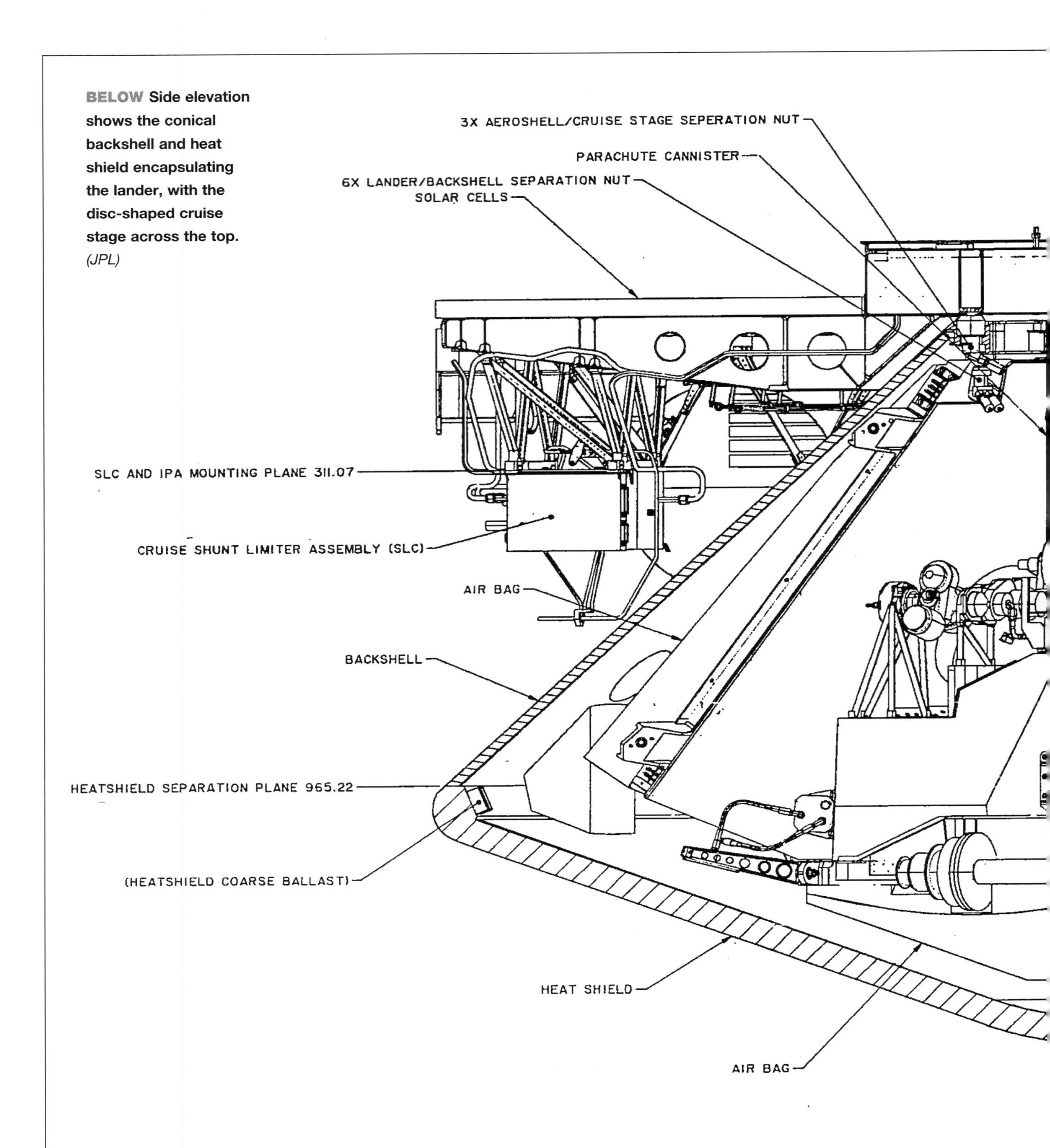

BELOW Side elevation shows the conical backshell and heat shield encapsulating the lander, with the disc-shaped cruise stage across the top. *(JPL)*

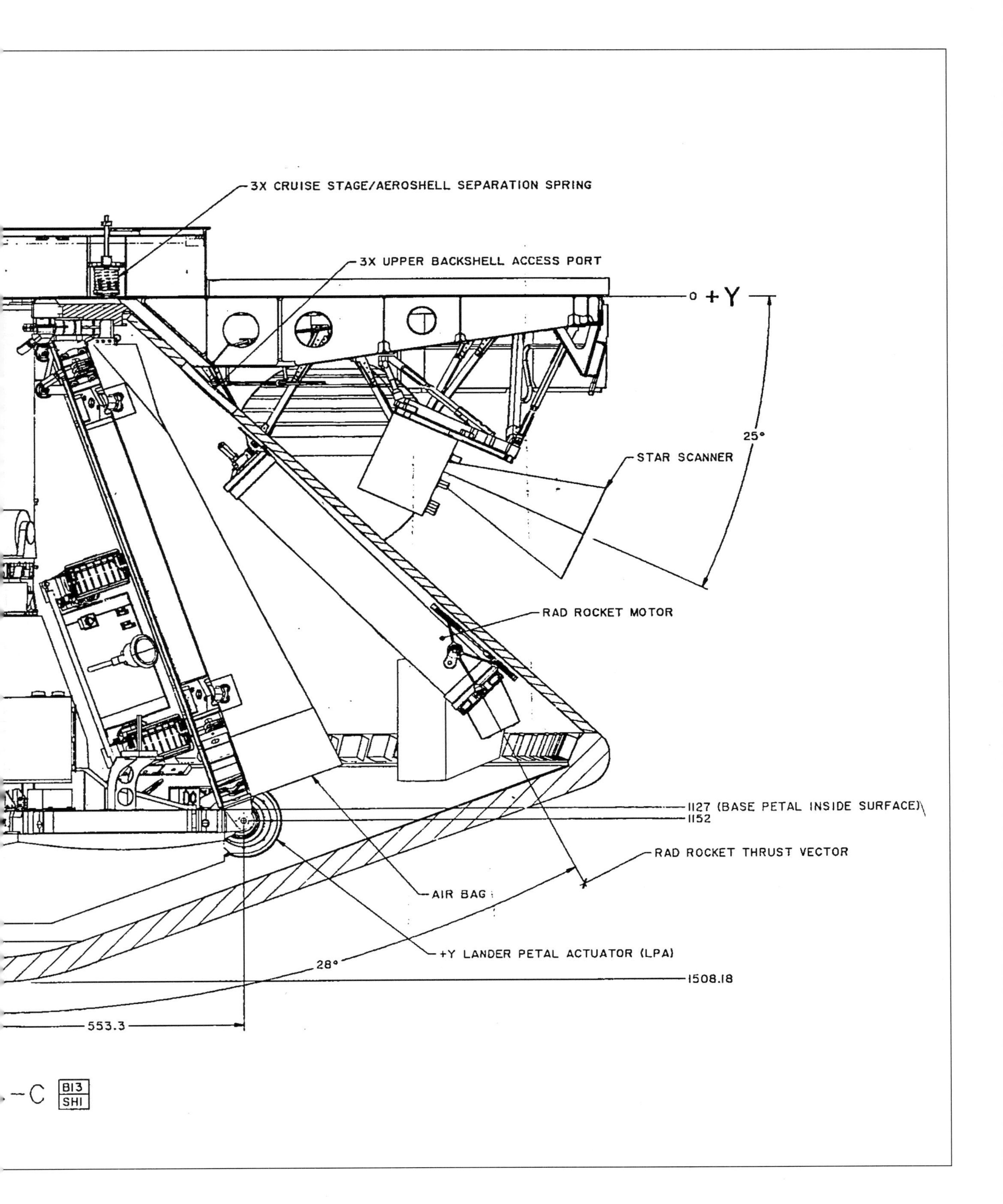
3X CRUISE STAGE/AEROSHELL SEPARATION SPRING
3X UPPER BACKSHELL ACCESS PORT
0 +Y
25°
STAR SCANNER
RAD ROCKET MOTOR
1127 (BASE PETAL INSIDE SURFACE)
1152
RAD ROCKET THRUST VECTOR
AIR BAG
+Y LANDER PETAL ACTUATOR (LPA)
28°
1508.18
553.3
C
B13
SH1

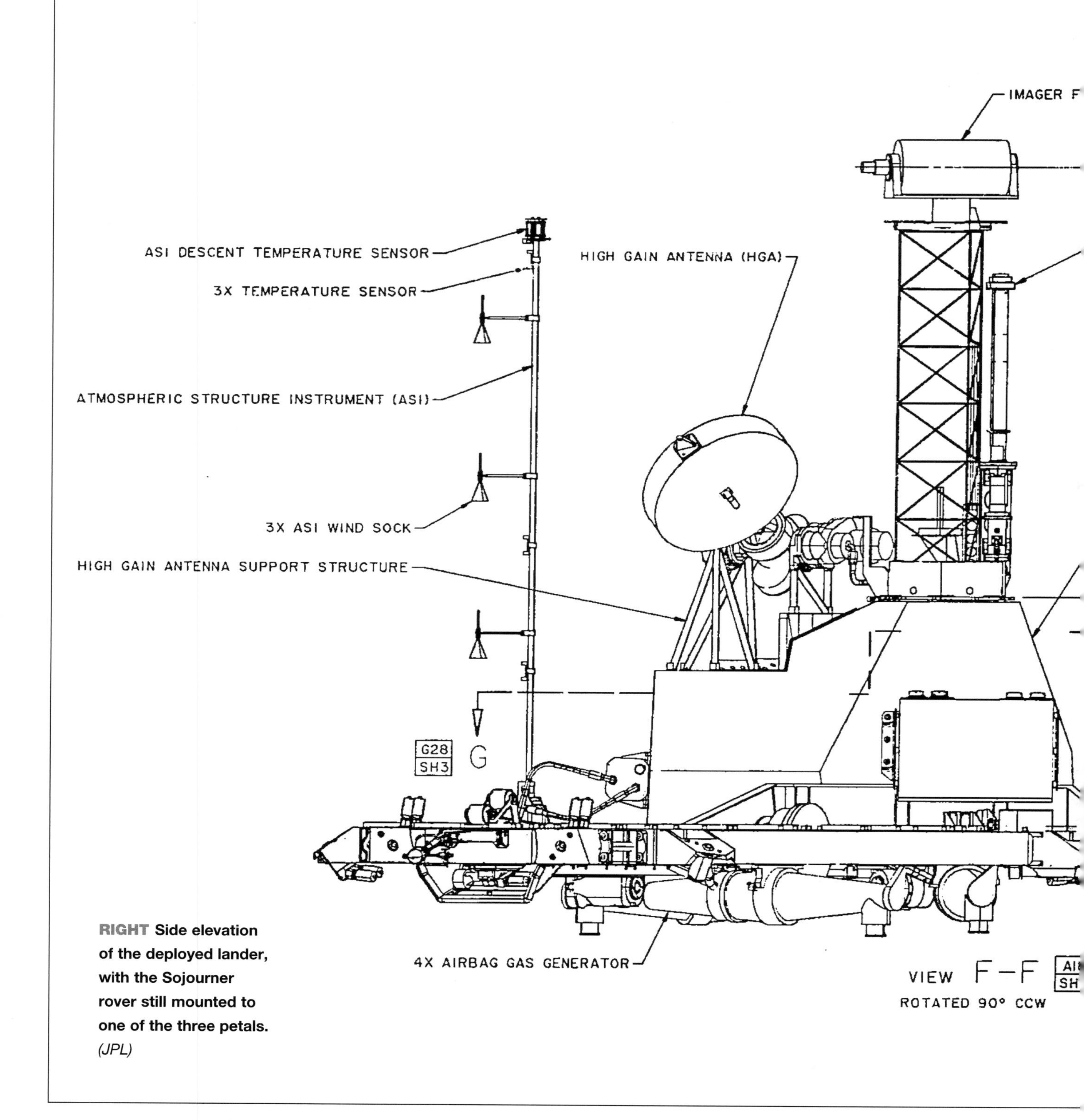

RIGHT Side elevation of the deployed lander, with the Sojourner rover still mounted to one of the three petals. *(JPL)*

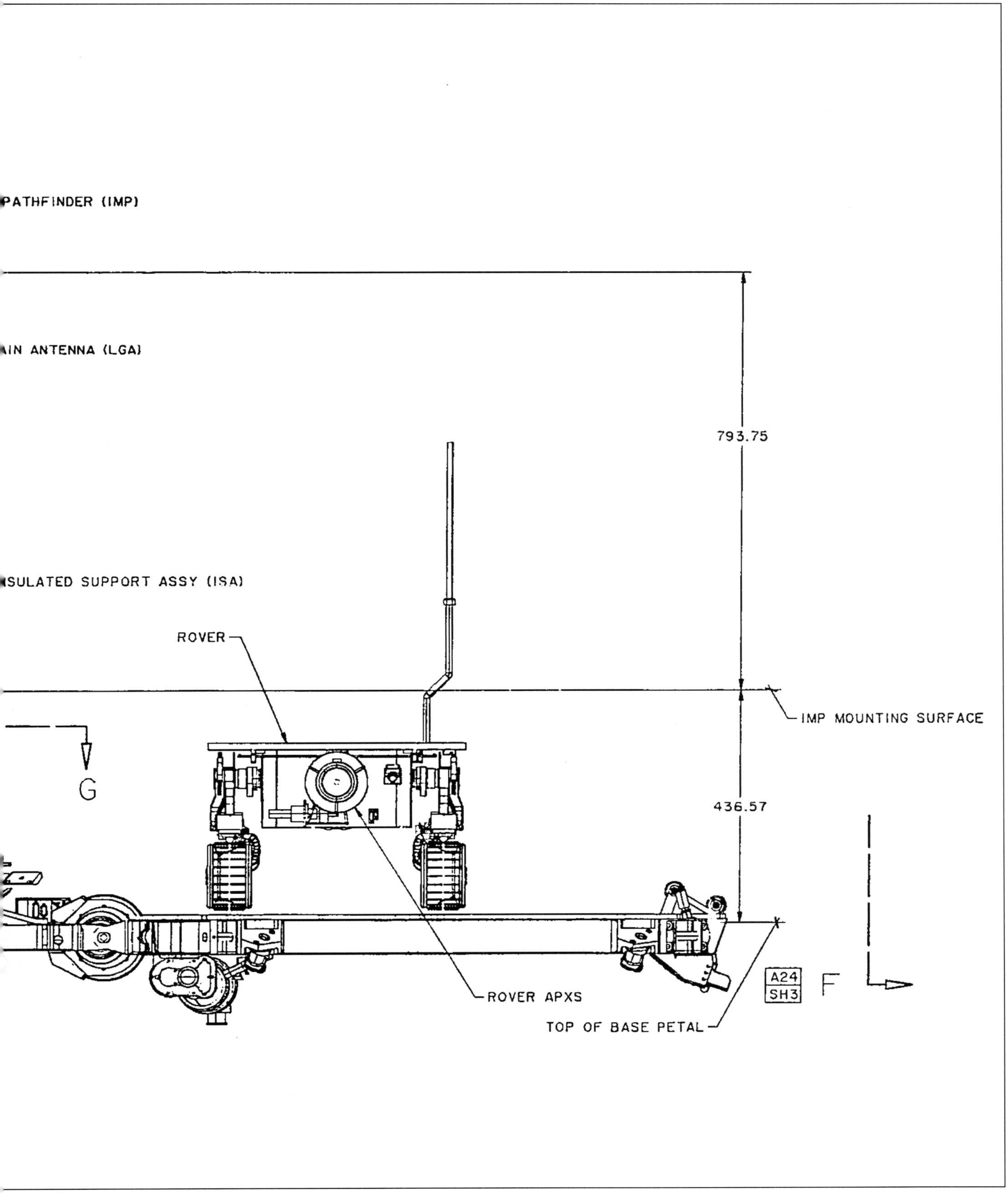

PATHFINDER (IMP)

IN ANTENNA (LGA)

793.75

SULATED SUPPORT ASSY (ISA)

ROVER

IMP MOUNTING SURFACE

G

436.57

A24
SH3

F

ROVER APXS

TOP OF BASE PETAL

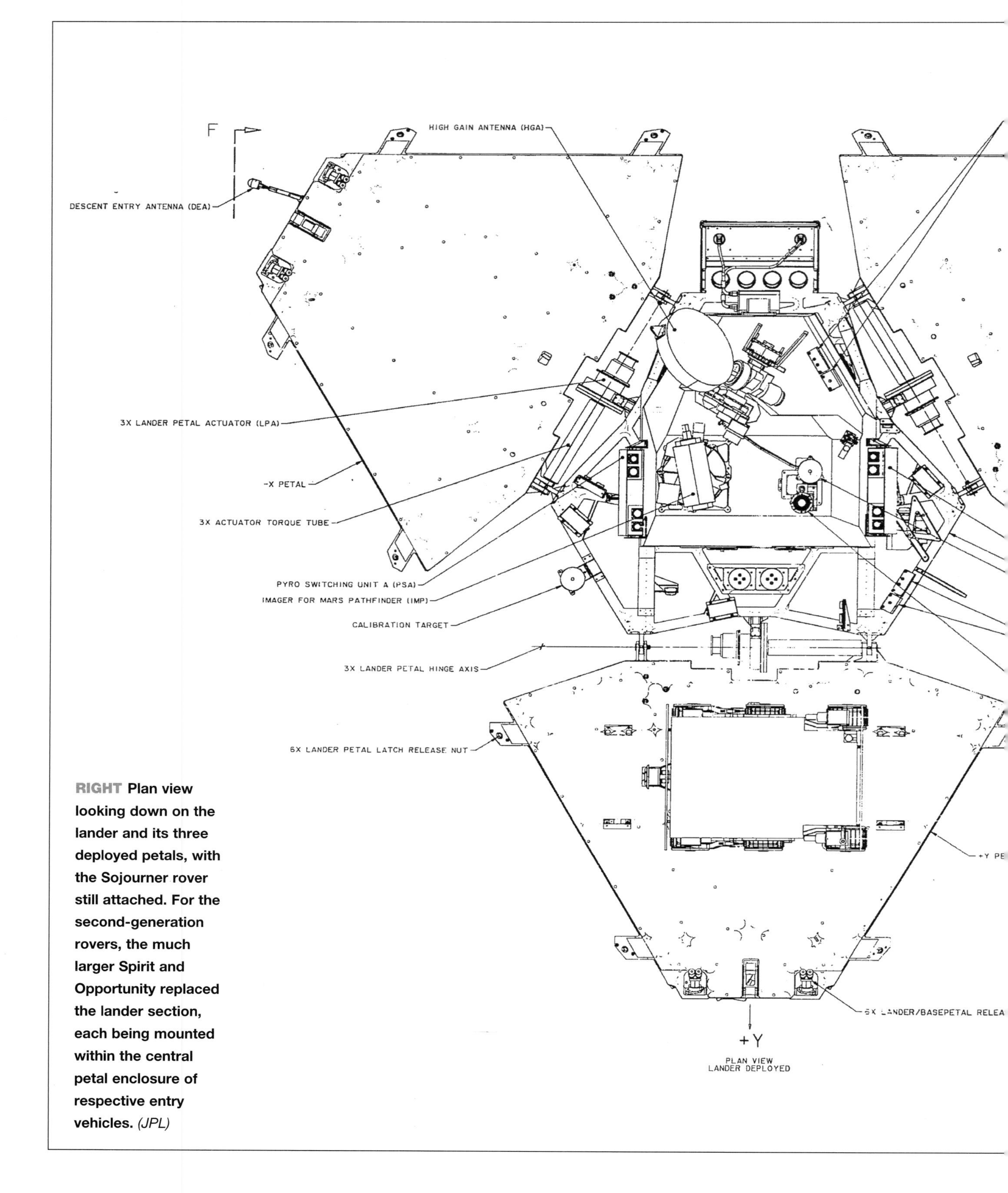

RIGHT Plan view looking down on the lander and its three deployed petals, with the Sojourner rover still attached. For the second-generation rovers, the much larger Spirit and Opportunity replaced the lander section, each being mounted within the central petal enclosure of respective entry vehicles. *(JPL)*

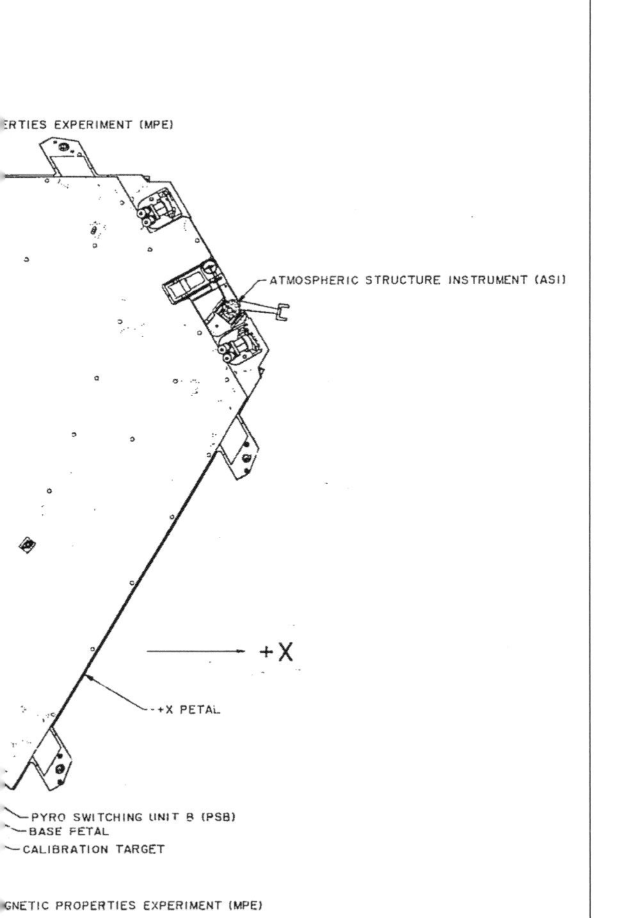

ABOVE An early view from the lander's mast camera before Sojourner 'stood up' from its stowed position. *(JPL)*

LEFT Freed from the lander, Sojourner drives a short distance to begin using its Alpha Particle X-ray Spectrometer on a rock. *(JPL)*

LEFT The rough surface demanded careful pre-programmed routes for the Sojourner rover, all but overwhelmed by the size of the rocks and obstacles. *(JPL)*

RIGHT A rock called Barnacle Bill was the first APXS target, the regions sensed being shown here in colour. *(JPL)*

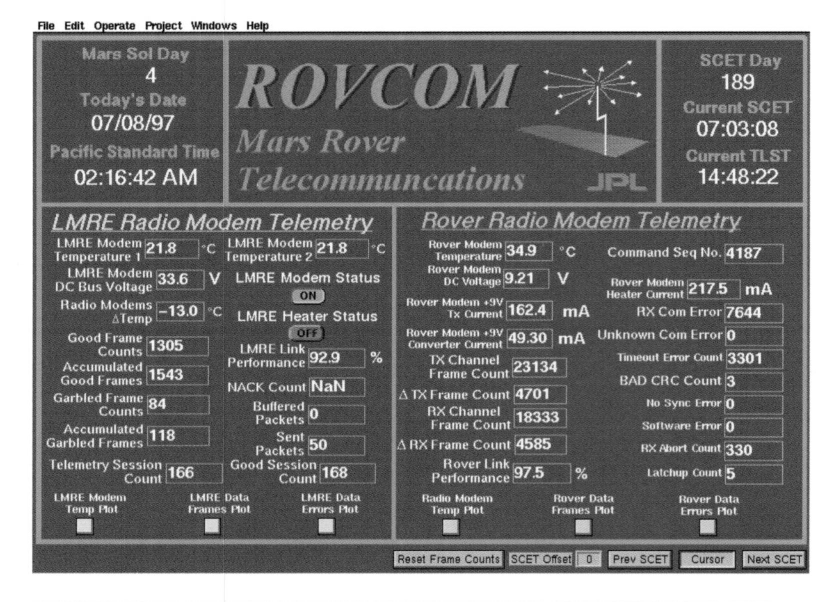

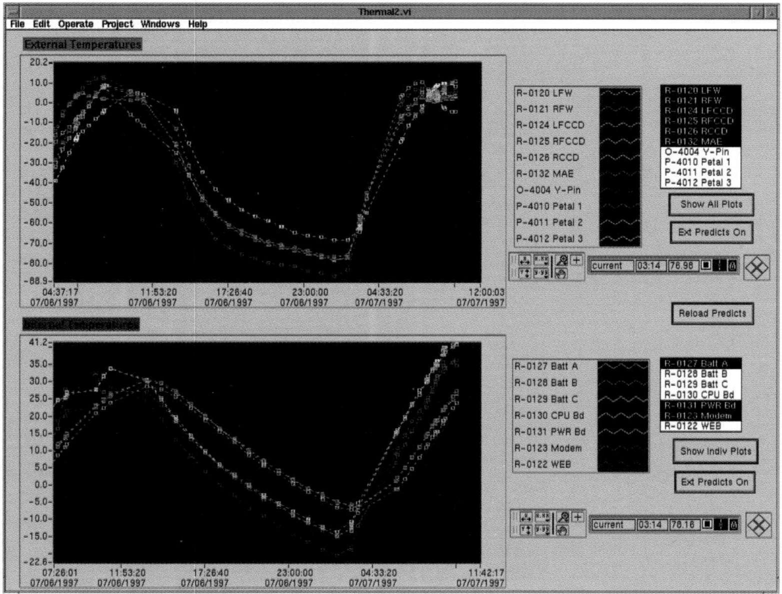

sensing instruments from orbit, such as that being conducted by Mars Global Surveyor.

Several engineering tests were undertaken using Sojourner, including terrain geometry analysis from images collected both by the rover and Pathfinder, soil mechanics tests from observed wheel sinkage, path reconstruction from dead-reckoning and track images, and vision sensor performance. One rover wheel carried different thicknesses of paint, abrasion of which indicated the amount of wear experienced by each wheel, a measure not readily discernible from images alone.

By the end of its life Pathfinder had transmitted 2.3 billion bits of data including in excess of 16,500 images, while Sojourner had captured 550 images and conducted 15 chemical analyses of rocks and soil at the

LEFT Pathfinder was the first Internet mission, the world wide web being an early vehicle for flight controllers eager to find a cheap and simple system for distributing spacecraft data to all the personnel involved. Here a sample page of the telecommunications readout for Sol 4 is displayed. *(David Baker)*

BELOW LEFT Thermal displays show external (top) and internal (bottom) rover temperatures across a day–night cycle, the Warm Electronics Box (WEB) keeping critical equipment above lower limits. *(JPL)*

BELOW Kinetic mapping of the rover on Sol 4 was crucial to designing traverse commands for Sojourner, with pitch and roll limits on angle of tilt across undulating terrain. *(David Baker)*

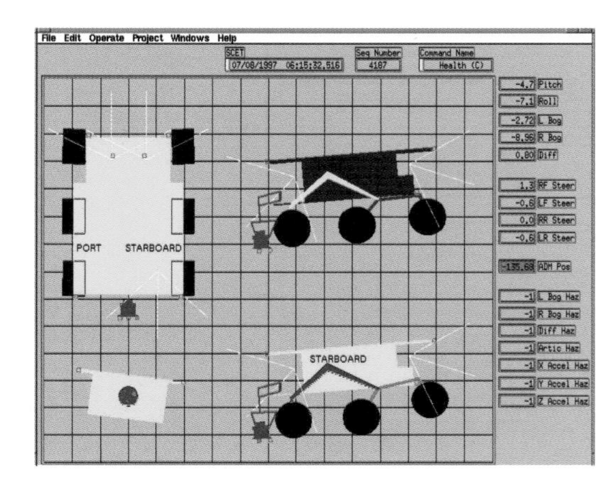

landing site. Shortly after landing the site had been named the Carl Sagan Memorial Station, after the noted planetary scientist Dr Carl Sagan who had done so much to popularise and explain the esoteric world of interplanetary space flight to millions of people through his TV programmes. Sagan had played a leading role in Mariner and Viking missions to Mars and the Voyager and Galileo missions to the outer planets before his death in December 1996.

Meanwhile, Mars Global Surveyor reached the end of its long journey to Mars on 12 September 1997. In a burn lasting 22 minutes, its rocket motor decelerated the spacecraft and placed it in a highly elliptical orbit, its weight now reduced to 1,691lb (767kg) after consuming rocket propellant. A lengthy period of aerobraking began during which the spacecraft shaved the outer atmosphere of Mars every time it reached the low point in its elliptical path, effectively lowering the high point of its orbit on each successive pass. Because the atmosphere of Mars is relatively thin, this aerobraking technique saved a large quantity of propellant that would have made the spacecraft much heavier had the technique not been feasible.

Only when the aerobraking phase had been completed in March 1999 did Global Surveyor begin its formal mission, although it had been operating its science instruments long before that. In its definitive orbital path the spacecraft would cross the equator of Mars each day during early afternoon, local time, and it would continue to provide excellent data, including stunning images of the surface, for almost a decade.

BELOW During three months of operation, Pathfinder lived up to the boldest of expectations, travelling a total of 328ft in 230 movements and exploring 820ft^2 of its landing site. Contact with the lander was lost on 27 September 1997 after it had operated for 12 times its design life. ***(JPL)***

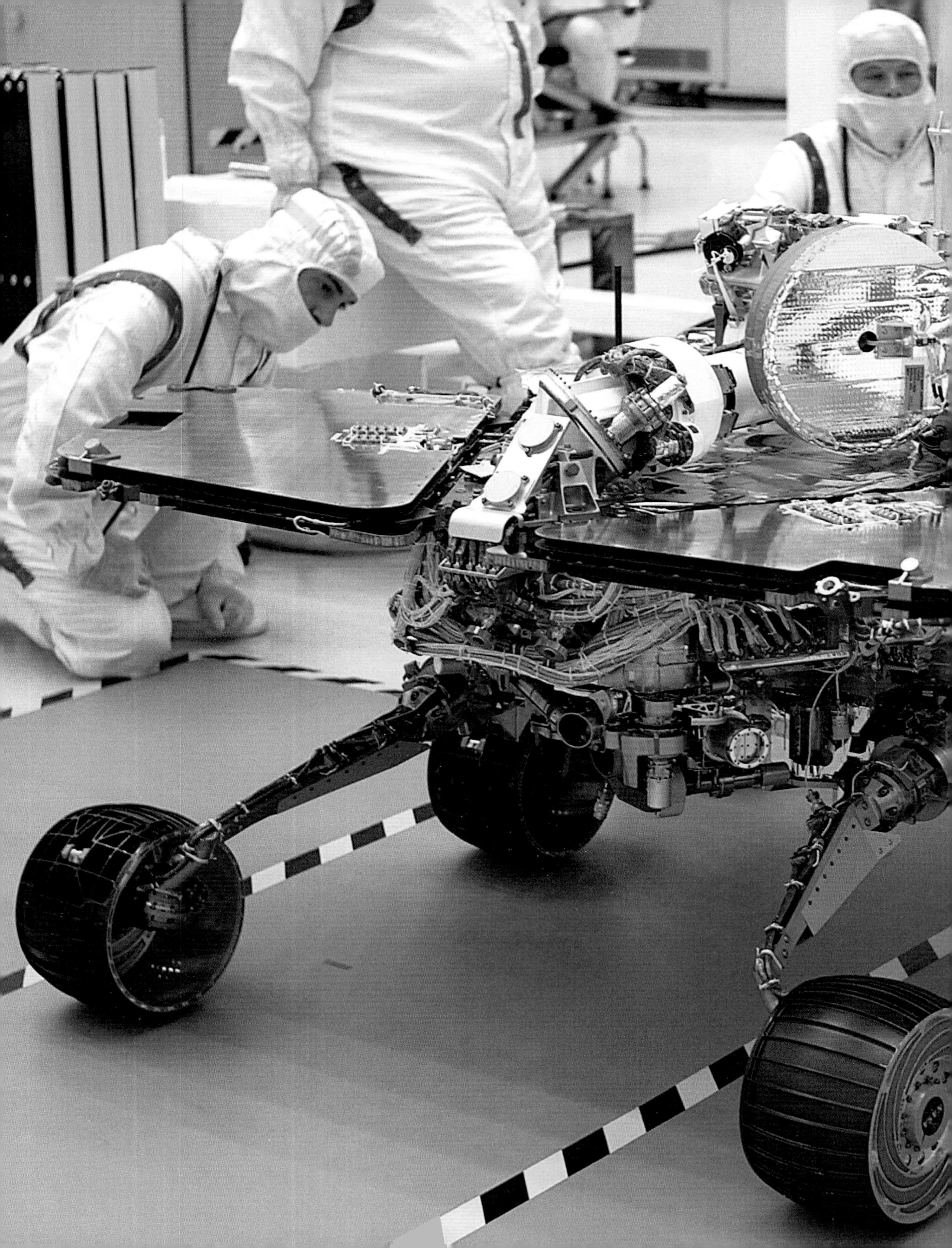

Chapter Three

Spirit and Opportunity

Two wheeled robots were selected at the dawn of a new century to pioneer a completely new way of exploring Mars. Each would land at a different site and each would conduct science that humans were unable to perform, on a planet now seen as having once been more Earth-like than previously imagined. They would outlast the wildest expectations of their designers.

OPPOSITE The rover as it would appear on the surface of Mars, its solar-cell panels fully open and deployed with the High-Gain Antenna erect and the camera mast folded. *(NASA)*

Disaster

Even before the launch of Pathfinder, the broad outline of a Mars programme anticipated the use of every launch window from 1998, beginning that year with two Surveyor-class missions – one to land on a polar ice sheet, the other to study the climate from orbit. In 2001 a Surveyor orbiter and a separately launched lander/rover were to be flown, a dual flight repeated in 2003 prior to a planned sample return mission in 2005. The sample-return flight was to land at one of the previous lander/rover sites and retrieve material its rover had collected for return to Earth. Those ambitious plans were to change when a double disaster struck NASA and undermined the entire Mars exploration programme.

The first of the Mars '98 missions, Mars Climate Orbiter (MCO), was launched on 11 December on a Type II trajectory that would

ABOVE Long before Pathfinder put Sojourner on the surface of Mars in 1997, engineers were thinking ahead to send a 'virtual' geologist with real science to perform. *(JPL)*

RIGHT Mars Climate Orbiter (MCO) was designed to economise on equipment by using off-the-shelf electronics, computer equipment and software. It had seemed to work for Pathfinder and Mars Global Surveyor, but doubts existed over the low-budget project long before the mission failed due to human error. *(NASA)*

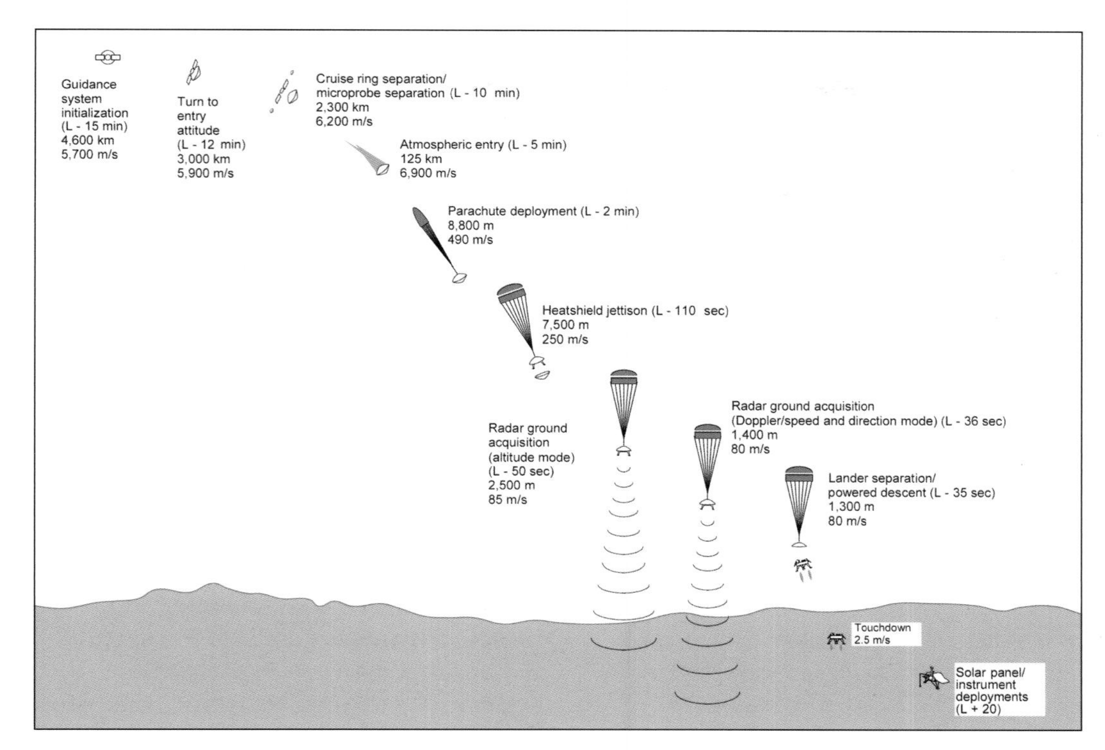

LEFT The descent profile for Mars Polar Lander (MPL) had more in common with Viking than with Pathfinder and its rover, parachutes slowing the descent to a speed of 175mph, where retro-rocket motors would take over. *(JPL)*

get it to its destination in late September 1999. MCO carried a 144lb (640N) thrust main rocket engine using a mixture of hydrazine and nitrogen tetroxide applying the power to decelerate into its initial aerobraking orbit of Mars. The second Mars '98 mission, Mars Polar Lander (MPL), was launched on 3 January 1999, on a Type II trajectory that would see it arrive at Mars after a flight of 11 months. Polar Lander would employ a combination of technologies from previous programmes, using aeroshell, heat shield, descent radar and parachutes similar to the Pathfinder spacecraft and descent engines like the Viking landers to power itself down to the surface on retro-rockets, but no airbags were used.

Instead of carrying a rover, the static lander had a sampling arm to dig into the surface near the South Pole in search of subsurface water. The arm would also deliver samples to the top of the lander where they would be transported internally to small ovens so that water and carbon dioxide could be driven off to determine the quantity of each.

Mars Climate Orbiter got there first, with a planned arrival date of 23 September, when it would be inserted into its initial, highly elliptical orbit. A final trajectory correction manoeuvre had been conducted on 8 September with calculations to fine-tune its low pass around the planet at an altitude of 90 miles (140km). These had been computed by the spacecraft manufacturer, Lockheed Martin, and sent to JPL for distribution to the tracking stations. As Climate Orbiter neared the planet, ground controllers noted that it was heading for a low pass of just 35 miles (57km) – too low to survive its brush with the atmosphere.

After it skimmed around the far side of Mars the spacecraft never reappeared on the nearer

BELOW MPL was designed to be the precursor to several Surveyor-series landers sufficiently flexible in design to eventually support a small roving vehicle like Sojourner. *(NASA)*

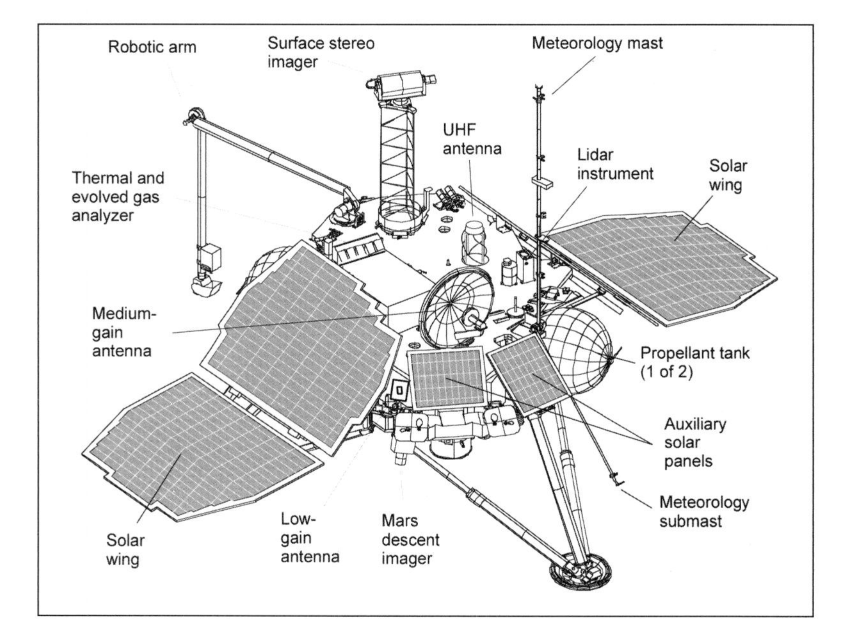

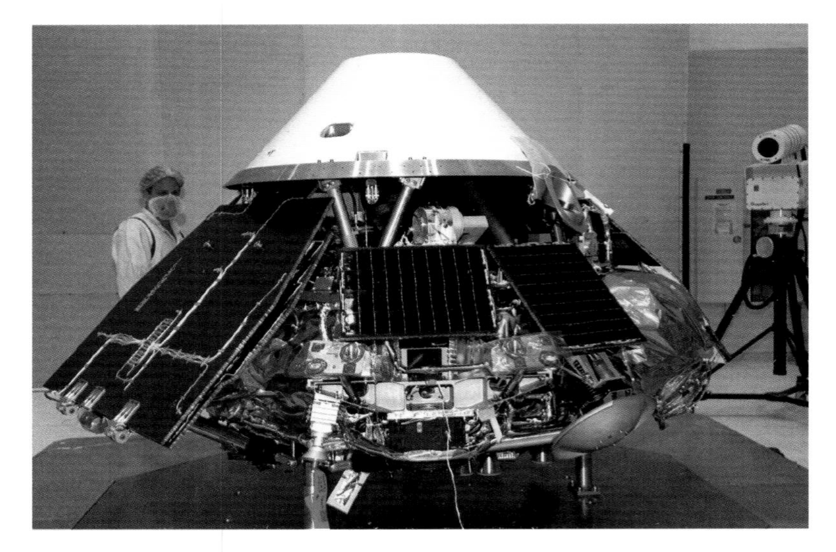

ABOVE Mars Polar Lander awaits its backshell and heat shield shortly before launch, which occurred in early 1999. *(KSC)*

side and nothing was heard from it again, its fate being to suffer a catastrophic breakup as it entered the denser layers of the Martian atmosphere. A review board found that the calculations conducted by Lockheed Martin had been made in units of pound-seconds instead of the metric units of Newton-seconds used by JPL, and in consequence the wrong commands had been sent to the spacecraft for the final trajectory correction burn, placing it far too close to the planet to survive.

Now it was time for Mars Polar Lander to go through a fast-paced sequence of events to put itself down on to the surface. As forecast, communication with MPL was severed at 06:39am local time at JPL on 3 December 1999, when the cruise stage separated and the spacecraft orientated itself for entry into the atmosphere, an event expected at 12:10pm. Signals were expected back from the Polar Lander at 12:39pm after it landed. Those signals never came. Another review board was convened, but without any record of the spacecraft's descent there was not much to go on. It did not take long to work out why it might have failed.

The most probable of the various possibilities, it was discovered, was in the simple action of deploying the landing legs prior to touchdown. In the automated sequence of events, the parachute would have deployed at a height of 6 miles (9km) and a velocity of 1,627ft/sec (496m/sec). Seven seconds later the heat shield would have jettisoned, followed 33 seconds after that by the landing legs deploying from their stowed position, 1min 30sec before touchdown.

In tests with backup hardware and the 2001 lander, engineers discovered that a fatal software flaw would have been triggered by the

RIGHT In deployed configuration in this artist's illustration, human error once again doomed the success of a mission to Mars, NASA's third to fail at or near the Red Planet. *(JPL)*

physical shock of the legs snapping out to their extended position, which had the same impact reading on the spaceframe as the spacecraft would sense when touching down on the surface. Some 47 seconds later the lander would have dropped free from the backshell and the descent motors ignited at a height of 5,340ft (1,628m) and a vertical descent rate of 256ft/sec (78m/sec).

Programmed to ignore software instructions on touchdown while the landing radar was on, when the radar was automatically switched off at a height of 131ft (40m) and a descent rate of 42ft/sec (13m/sec) the touchdown-sensor logic kicked in and immediately interpreted the stored signal as indicating the spacecraft was already on the ground. So it immediately cut off the descent motors and the spacecraft fell from that height, impacting the surface at 72ft/sec (22m/sec), or 49mph, far above the planned 7.8ft/sec (2.4m/sec), or 5mph. It could not possibly have survived the impact.

Recovery

In April 2000 NASA set up a new Mars Program Office, its point of contact at JPL for all Mars missions and architecture studies, headed up by Dr Firouz Naderi, a veteran of complex and advanced space science missions. The decision was quickly made to fly the planned Surveyor orbiter at the 2001 launch window but cancel the lander/rover that had previously been twinned with it. Resurrecting the 2001 Surveyor orbiter, NASA renamed it Odyssey, a tribute to Arthur C. Clarke's book *2001: A Space Odyssey*. But plans for a lander/rover would have to wait until the 2003 launch window.

Mars Odyssey was launched on 7 April 2001 on a fast Type I trajectory that would take approximately 200 days. The spacecraft weighed 1,609lb (729.7kg), which included 779lb (353.4kg) of propellant to insert itself into a preliminary orbit about Mars and a suite of science instruments with a weight of 98lb (44.5kg). The camera carried by Odyssey had both optical and infrared capability, allowing scientists to get a thermal image of the surface as well as a visual picture, showing surface features down to about 59ft (18m) in size.

ABOVE Recovering from the failed 1999 orbiter and lander missions, NASA returned to Mars in 2001 with Odyssey, weighing 1,609lb at launch, half of which was propellant for easing into an aerobraking orbit, seen here. *(JPL)*

BELOW Designed to conduct a full survey of the distribution of minerals across the surface of Mars, in addition to an imaging system Odyssey carried a gamma ray spectrometer on the end of a long boom with which it would return data on up to 20 chemical elements. *(JPL)*

Odyssey arrived at Mars on 23 October 2001 and successfully decelerated into an orbit about the planet. From there a protracted series of aerobraking manoeuvres were completed in January 2002 and operations commenced on 19 February. Still operational, Mars Global Surveyor had completed its primary mission a few weeks earlier and was now embarked upon an extended operational period that would see it continue to function until 2006 when contact was lost due to human error in sending commands. Meanwhile, Odyssey began a new survey of the planet that would continue well beyond the arrival of the Curiosity rover in August 2012, making it the longest-lived orbiter around another world.

A year before the launch of Odyssey, in 2000 NASA moved swiftly toward a decision about the next stage in the exploration of Mars and considered either a scientific orbiter or a large roving vehicle for launch in the May 2003 window. Two teams, one at JPL and the other at Lockheed Martin, conducted independent two-month studies to define the concepts and present them for top-level decision. The two proposals were submitted to Scott Hubbard, the Mars Program Director at NASA headquarters, for a decision by Ed Weiler, the Associate Administrator for Space Science. Lockheed Martin was responsible for the Mars Science Orbiter (MSO) concept while JPL had put together a plan to fly an advanced rover it had already been developing.

RIGHT Steve Squyres was almost single-handedly the most influential person in getting the Mars Exploration Rover project approved ahead of a competing contender for the 2003 launch slot in the form of another orbiter. *(David Baker)*

MSO originated in a concept dubbed Gavsat, named after Thomas R. Gavin, director of JPL's Space Science Flight Projects Directorate and recipient of numerous NASA awards. He wanted a safe project to reassure doubters that the Mars '98 failures were aberrations and not typical. Politics played a big hand in what NASA got from Congress and what the public wanted the government to fund. Mars Odyssey had been one of Gavin's projects and Gavsat was his concept for a bigger orbiter with the latest technologies in cameras and science instruments. Officially known as Mars Science Orbiter, it was the least risky proposal – on paper – but only because Odyssey had yet to be launched and nobody could be certain it would be successful. But other people had different ideas.

Mark Adler was a genius with software programs and was the inventor and author of several compression libraries and network graphics, including the PNG image format. A few years earlier he had been the manager of the Cassini-Huygens mission to Saturn and from 1996 to 1998 he was the architect of future Mars missions. Now he was pushing an idea that would put wheels back on the Red Planet. Adler wanted to fast-track a new lander mission by not only adopting the same technique as Pathfinder, but bolting a new lander into the existing aeroshell developed for that highly successful spacecraft, saving money and development time.

Next up was Steve Squyres, a graduate of Cornell University, where he obtained his PhD. Throughout the 1990s he had nursed hopes of putting instruments on Mars and worked hard lobbying for a roving vehicle to conduct geological work across the surface. Determined to get to run the science mission of this still hypothetical lander, Steve would be a key player in the next phase of decision-making over the 2003 launch window, and his affection for placing rovers on Mars went back several years.

But he needed someone to help put a formal proposal to senior management, and found him in a seasoned veteran of former missions. Barry Goldstein began working at JPL in 1982 as a flight designer for the attitude systems on the Galileo spacecraft to Jupiter and then became lead system engineer on a payload for the ill-fated Mars Polar Lander. He was an ideal choice to pull together all the elements of a new mission and Steve Squyres invited him in to manage the lander proposal up the chain to higher authority at JPL.

The proposal was based on the existing Athena rover design, a concept that had re-emerged during planning for the 2001 and 2003 lander/rovers leading up to a full sample-return mission around 2005. For these missions the lander would carry the 165lb (75kg) Athena down to the surface using a combination of atmospheric braking, parachutes and retro-rockets. But it was too big to use the airbags method selected for Pathfinder, and instead would decelerate using braking rockets on the underside of the lander much like the failed Mars Polar Lander of 1998. The lander would unfurl two circular discs of solar cells and the Athena rover would be carried on top and deployed down a ramp to the surface. Its function was to conduct a geological survey using sophisticated science instruments.

Athena was calculated to survive 90 Sols (Mars days) at the surface after landing at a mid-latitude site. It was designed to carry panoramic cameras (Pancams), a Miniature Thermal Emissions Spectrometer (Mini-TES), a miniature coring device, an Alpha-Proton X-ray Spectrometer (APXS), a Mossbauer Spectrometer, a Raman Spectrometer and a micro-imager. The Pancams were two separate high-resolution stereo cameras operating in slightly different wavelengths attached 30cm apart on the T-bar at the top of the mast.

The Mini-TES could detect infrared emissions from objects at some distance and determine their mineral composition. With the ability to 'see' through obscuring dust, the thermal emissions could differentiate between carbonates, silicates, organic molecules and minerals of the type formed in water. The

ABOVE The new rovers were designed around remote and contact science tasks, the former including a Mini-TES (Miniature Thermal Emissions Spectrometer) capable of detecting different chemicals and minerals formed in water by observing thermal emissions at a distance. *(JPL)*

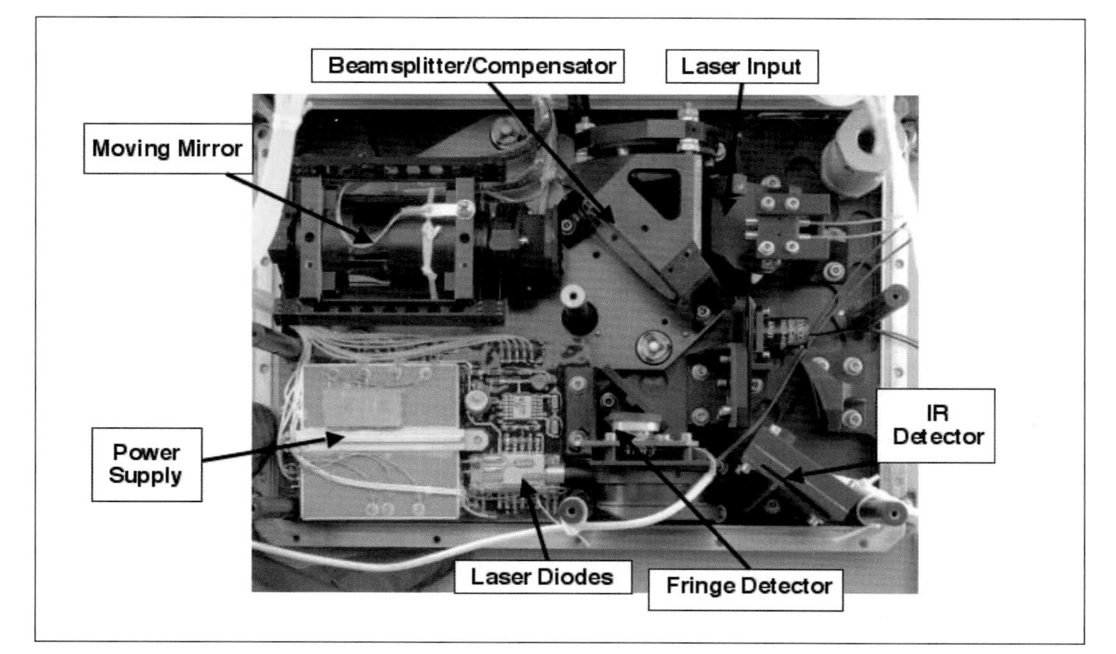

LEFT The optics bench for Mini-TES included the laser, beam-splitter and the power source for what would turn into one of the most useful instruments carried by the rovers contributed by German scientists. *(JPL)*

APXS was similar to the instrument carried to the surface of Mars by Sojourner in 1997 and would determine the abundance of rock-forming surface materials.

In addition to detecting magnetic properties of surface materials, the Mossbauer Spectrometer was primarily to determine the quantity and abundance of iron-bearing minerals. To grind away the surface of a rock and allow a micro-imager and the other instruments to gain access to subsurface structure, a Rock Abrasion Tool (RAT) would grind out a circular section 2in (5cm) in diameter exposing internal materials to a depth of 0.2in (5mm). The Raman Spectrometer was proposed as a means of determining the chemical bonds and the symmetry of molecules that made up various structures and materials at the surface.

The origin of Athena went back to a meeting called by Steve Squyres in January 1996 to discuss the possibility of a Mars lander for a mission to launch in 2001, but that had been cancelled. The 'new' Athena proposed for launch in 2003 would weigh 286lb (130kg) and adopt the same delivery concept as Mars Pathfinder, but the lander section would serve merely as a base platform.

This rover would be fully autonomous, cleared to drive up to 320ft (100m) per day and traverse difficult terrain. It was to land at a site optimised for mineralogical and geochemical analysis in an area believed to have been covered in water at some point in the evolution of the planet. It would be designed to carry a wide range of tools, instruments and imaging cameras for sophisticated robotic exploration – all encased within the same size lander as was designed for Pathfinder.

RIGHT **Software guru and the inventor of several data compression technologies, Mark Adler designed the Mars Exploration Rover mission architecture, pushing for use of Pathfinder hardware but with a new and more capable rover. He would become mission manager for Spirit.** *(JPL)*

When, on 26 April 2000, Mark Adler presented the lander/rover proposal – now named Mars Geologist Pathfinder (MGP) – to John Casani for a top-level overview, he secured interest that seemed to place it ahead of its rival Mars Science Orbiter. Casani had the job of examining all the various mission proposals and rating their success, as well as assessing their risk. Senior management favoured Athena and the MGP mission. The next step was to sell it to NASA management, a hard sell because the rival team was equally determined to get their Mars Science Orbiter approved. For that contest, JPL changed the name yet again to Mars Mobile Pathfinder (MMP). Over two days at JPL beginning on 3 May 2000, the competing proposals were brought for evaluation before a team from NASA headquarters in Washington led by Ed Weiler. Three proposals were under way: the orbiter, the rover, and a repeat of the failed Mars Polar Lander – with modifications.

In some perverse way, the failures of Mars Climate Orbiter, built by Martin Marietta, and Mars Polar Lander, built by Lockheed Martin, had shifted the comfort zone back to an in-house effort. Politically, pulling back from low-cost, two-at-a-time missions to a more conservative, properly funded, single mission at any given launch opportunity with a device built in-house was probably more palatable to politicians who would approve or deny funding for the project.

In the unpredictable and oft misplaced logic of politics and government-run space programmes, NASA would request approval of the government (the President), who would then agree a budget and propose it for funding to a cynical group of politicians in Congress. As the NASA hierarchy and the planetary wannabes paraded back into their meetings at JPL on the second day of the reviews, everyone knew 4 May was make or break. As they listened again to the presentations and discussed subtle points of each, it was clear that the Athena-based rover proposal was gaining ground. It just seemed to make so much more sense.

Odyssey was now on its way to Mars, and the previous orbiter, Mars Global Surveyor, was performing well. Why fly yet another orbiter when it would be more sensible to wait at least one more launch window for results from what would then be two Mars orbiters in play before sending a third? It was better, surely, to use that slot to capitalise on Sojourner and push forward with surface science exploiting a truly flexible and well-equipped rover to do some real science?

JPL needed a bold new initiative more than anything. As Ed Weiler roamed the laboratories and the departments at JPL, chatting with the scientists and the engineers, the deep desire to press ahead and be given a strong, bold and innovative challenge was palpable. And that was the impression he took back with him to Washington. Ed Stone was convinced that the rover mission had it sewn up and the following day authorised a project team to prepare such a mission. It would be managed by Pete Theisinger, another veteran of past glories, with Richard Cook, another seasoned Mars scientist, as flight systems manager.

On 13 May NASA formally announced that both MSO and MMP were being considered, the reworked Mars Polar Lander bid had been dropped, and that each project would receive $4 million to come up with definitive plans by July. Everyone examined areas where weight and/or cost savings could be made. The definitive Athena rover forged in the critical two months before the final run-off between contenders managed to retain almost all its scientific experiments except the Raman Spectrometer and the mini-corer, and there was a serious problem with keeping the Mossbauer Spectrometer. It was deemed to be valuable as part of the science payload, and an integral part of the final proposal to impress the leadership with the science Athena could do. So Steve Squyres got Rudy Rieder from Germany to donate a specially refined model that the Europeans had developed for their Rosetta comet mission. The one carried by Sojourner was not as good as the technology could now produce.

By the time the final showdown was staged at NASA headquarters in Washington DC, it had been firmly agreed to shoehorn the rover into the same aeroshell as had been developed for Pathfinder, and that a direct entry would be flown using the same technology for arriving at the surface, including airbags. When it was presented on 13 July, the first of a two-day series of briefings and analysis, the name had changed again. Now it was known as the Mars Geological Rover (MGR), to emphasise the science it was expected to carry out. And that was appropriate, because the first day of the briefings was spent examining the science that could be conducted by an orbiter and by a rover. The second day considered the technical and engineering issues.

While the orbiter was certainly less risky, requiring relatively benign and tested techniques to achieve its goal, the lander ran a much higher probability of failure, if only because it was a more complex engineering challenge. There was an additional reason, however, for choosing the rover. It just so happened that the gymnastics of orbital dynamics made 2003 the year in which Mars and Earth would be closer together than at any other launch window in an 18-year period, allowing the launch vehicle to carry just that little bit of extra payload.

The final interim decision was made during a teleconference between NASA and JPL on Tuesday 18 July. As far as Ed Weiler was concerned, Athena was going to Mars! All it needed now was the approval of the NASA Administrator Dan Goldin, and the presentation to him would be made on the Friday of that week. Goldin made it known that he had expected the orbiter to be selected and was going to need a lot of persuasion to change his mind. Within little more than an hour after meeting with Goldin on 21 July, the presentation was over and Ed Weiler had his rover mission – except for one thing.

Concerned at the high risk involved, and not wanting to provoke the ire of politicians lambasting NASA for wasting another launch window, Goldin decided to go for two rovers instead of one. If one failed, the second would save the day. And if both succeeded, NASA would have totally vindicated itself. But he needed to know how much extra that would cost and within one hour the team came back with an answer: $440 million for one; $665 million for two.

Goldin took the options up to the White House, which refused to come up with the extra money. It was up to Goldin, they said, to find the extra $225 million out of NASA's existing budget. The agency was boxed in and Goldin was resolute. He had already signed up to the rover mission but he would not risk everything on a lone spacecraft.

On the morning of Wednesday 26 July the team got the word in a telephone call from headquarters to say that NASA would find the money come what may, and a press release to that effect was released the following day. And as if it had not had enough name changes, the Athena rover programme would now be known as the Mars Exploration Rover (MER) mission. Less than three years remained to get the two spacecraft to the launch pad.

The second-generation rovers

When the MER team got their instructions to build not one rover but two it posed more problems than it solved. Politically, it was wise to have a second rover fly to Mars, but spacecraft are not assembled on a production line and each is unique. From the time components start to come together in a clean-room where dust and microbes have no hiding place, until the end of their mission at assigned destinations, two spacecraft built to identical specifications will each perform very differently to the other; which means that technicians and engineers have to learn to map subtle variations in response to signals, different reactions to computer loads, and operating nuances – which means twice the work of building just one.

Moreover, with two launch vehicles and two flights under way simultaneously, the workload for tracking and monitoring facilities is doubled, as is the number of people needed to run the computations, deliver new software packages and run tests to make sure that only the correct commands get sent to the spacecraft. And, because two virtually identical spacecraft would be operating at the surface of Mars at the same time, the landing sites would have to be spaced so that tracking stations could phase communications as Mars rotated.

Ideally, landing sites should be on opposite sides of the planet so that when one spacecraft was around on the far side to Earth and resting in darkness, the other could be operating with full sunlight maintaining electrical power. And having the two spaced in distance would mean the operational planning cycles could be phased in time as much as 12 hours apart. Another aspect involved work/rest cycles for the ground controllers and the mission planning teams. The 39 minutes extra in each Mars day, or Sol, would play havoc with the mission control teams, requiring them to keep Mars time and gradually migrate their working day by 39 minutes every 24 hours. It was nice to have twice as much science to do, but it had its price.

The specification for the two MER rovers was tight. Much depended upon the technology. Sojourner had operated for a few weeks, but, being a technology demonstrator rather than an operational rover for exploration, that was acceptable. The two MER Athena rovers would be required to work hard and to deliver real science, so the engineers at JPL were tasked to come up with a minimum operating time that could be built into the NASA requirements document that would, at the end of the mission, determine whether it had matched promises and expectations. This document was an important yardstick by which the performance of the team and the credibility of individuals

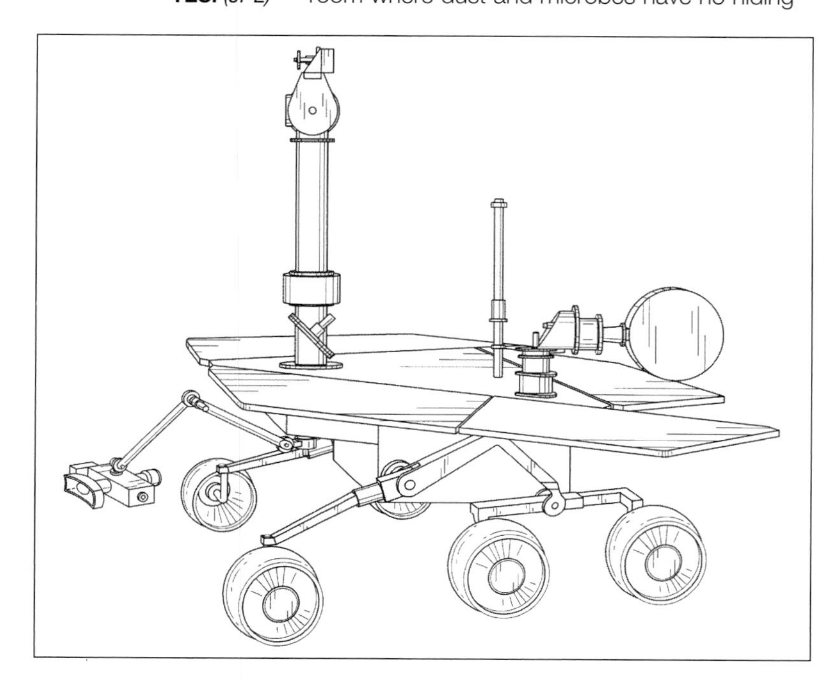

BELOW The basic Athena concept incorporated the mobility system pioneered by Sojourner with fold-out solar array panels, a mast for remote science including imaging and infrared spectroscopy, and a boom for contact science with the Mini-TES. *(JPL)*

LEFT Athena was designed so that all the sensitive electronics would be inside a Warm Electronics Box with the ability to send data direct to Earth as well as through orbiters such as Mars Global Surveyor and Odyssey. *(JPL)*

would be measured. Too high an expectation and it could be difficult to live up to.

The baseline figures agreed upon established for each rover a minimum operating period of 90 sols on Mars, just over 92 Earth days, with a total driving distance of 1,970ft (600m). In that period scientists would explore the region around the landing site with real-time planning depending upon the features observed after touchdown. For many, especially the scientists, this was simply not enough. With strong, robust little rovers festooned with science instruments and equipment, the urge to build in more 'life' than necessary to meet the requirement was superfluous and a potential hazard to achieving the basic mission. But that didn't stop anyone from trying to build it much better than they needed to – just in case they lasted longer than three months in the hostile, freezing environment of Mars.

A part of that was to ensure that each rover had as many solar cells as possible. Electrical power is lifeblood to a rover on Mars, without which it will not have energy to communicate or power any of its systems. And experience with Pathfinder and Sojourner had shown that the blanketing effects of dust and wind, albeit tossed around in a very thin atmosphere, could seriously degrade the potential of the photo-voltaic cells to produce electricity. So a push by the scientists for the engineers to get as much solar power on board as possible was a bid to give each rover a longer life on the surface.

The cells would be applied to flat panels that would unfold when the rover emerged from its lander section, each string of 16 cells wired into the electrical circuits. The number of strings that could be carried on the panels would not only govern the electrical loads the equipment could draw upon, but would also determine the reserve power the rovers would have as dust began to blanket the glass-like surfaces. The basic design involved a hexagonal array consisting of three foldout panels, each one attached to the long side of a fixed solar-cell panel mounted flat on the top of the rover.

BELOW The traction system was a legacy design from Sojourner, retaining the rocker-bogie but with metal wheels incorporating short spikes for grip. *(JPL)*

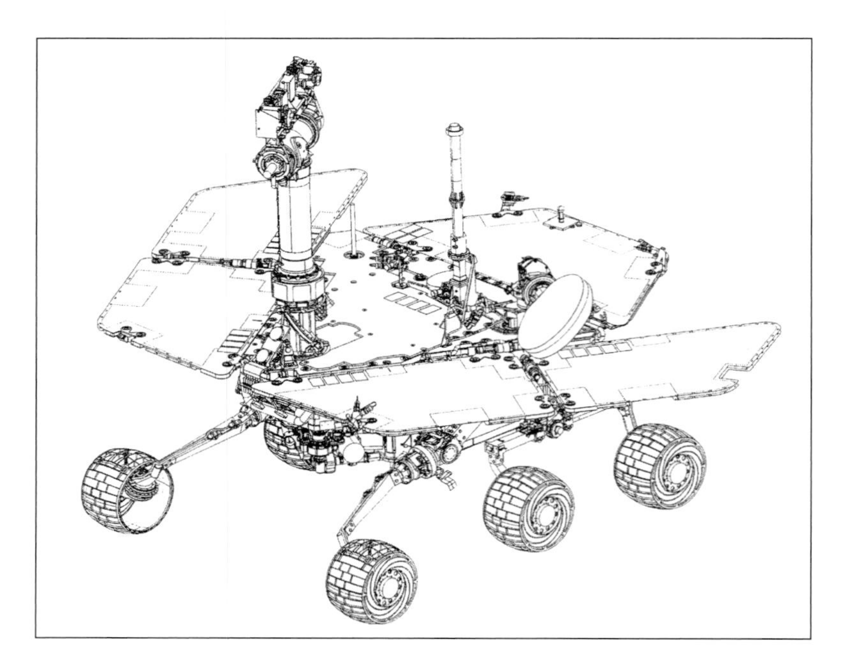

ABOVE The definitive design concept of Athena morphed into Mars Exploration Rover with refinements and a change of wheel configuration. *(JPL)*

Because the rover would be folded up inside its triangular pyramid-shaped lander, the foldout panels could only be as big as the remaining space inside the enclosure.

To support a 90-Sol mission and have some energy in reserve, scientist Steve Squyres wanted to see at least 36 strings of cells. The first design gave him 26 strings at most. With redesign brought on by sheer pressure of necessity, and a desire to give each rover the best chance as possible for at the very least 90 Sols on Mars, the solar panel 'wings' were arranged to fit 34 strings – just good enough. But weight was a problem and the additional solar array segments, plus several other pieces of equipment, were evaluated to see if they could be removed. Because the rovers were to use the same-size lander and aeroshell configuration, weight and size were a critical element in not only packaging the rovers into the same-size structure but also achieving it within a weight that would not exceed the capacity of the Delta rocket.

A Critical Design Review, or CDR, was held in August 2001 with the aim of finalising the design and fixing the specification prior to building the two rovers. It was becoming apparent that simply using the same lander as Sojourner would not work. The MER vehicles were much bigger, more sophisticated, and would carry all the communications equipment and instruments on to the surface. In all essential aspects, the lander would merely be a platform from which the rovers could dismount and go about their work. That additional responsibility placed on the rover design added weight and, potentially, cost to the programme.

Sojourner had been a technology demonstrator carried on one of the lander's three deployable petals – a lander that was, by definition, itself a demonstration of a completely new way of putting a working package down safely on the surface of Mars. In contrast, each MER rover would be the object of the mission itself, an operational vehicle capable of doing complex and worthwhile science, mounted on the top of the lander platform and not as an appendage. But, while the rover design was growing in complexity and girth, the lander package remained resolutely the size it had been at the start of the programme.

Dismay at the squeeze placed upon the programme management by the CDR spilled over to a day of reckoning steered by Steve Cook, flight system manager. Despite supreme effort, where several experiments and even the additional solar panel segments were considered candidates for a weight reduction effort, everything that had been designed-in remained aboard the rovers. Weight was shaved gram by gram from every conceivable part of the spacecraft that could offer some contribution to the drive to keep the total mass below the redline limit, and that was a remarkable achievement in itself.

By October 2001 it was clear that weight was not the only thing going up: the estimated cost of the missions from start to finish was also increasing. Working to a 'ceiling' of $688 million set when the two rovers had been adopted, estimates calculated by the project team now had that rising to at least $748 million. A NASA review body moved in and said that in reality the MER missions were likely to cost a lot more, but Ed Weiler, head of planetary science at headquarters, allowed the ceiling to rise to $746 million. NASA had looked at cancelling the second rover but by now the team had become used to having the two vehicles to work with and were horrified at the thought of returning to the one-shot flight. So two it was – unless there were further cost increases, in which case one would have to go.

For readers unfamiliar with the tight constraints placed on spacecraft design engineers, the search for a few grams in weight savings may seem obsessive, even ridiculous. But it is not. During the search for anything that could shave weight from a bloated design, engineers had removed a tiny camera called Suncam, which was added initially to take pictures of the Sun to help the rover navigate around on the surface. It weighed only 300g but it was a luxury whose function could be carried out by the two panoramic cameras (Pancams) instead, so it came off. Now that weight would save a critical function missing from the rover: an ability to calculate the sideways motion seconds before impact.

Tests with the airbag configuration revealed a susceptibility to abrasive tearing of the thin fabric lobes in a high lateral, or sideways, motion. During the final few seconds of descent the inflated airbag cluster had to be dropped straight down, with no lateral motion such as might be caused by high winds close to the ground. A Transverse Impulse Rocket System (TIRS) was adopted, comprising thrusters that could fire to stop the airbag cluster from sideways motion, stabilising it to a true vertical descent.

To calculate the amount of drift in the descending spacecraft, a camera was added to the base of the lander. When the radar had acquired the surface the camera would take three pictures four seconds apart to compare ground features and display the degree of drift. From this the computer could calculate if, and for how long, the TIRS rockets should be fired to compensate for sideways movement. Because the Suncam had been taken out, and other weight savings made, there was capacity for the new and more vital optical device. Called the Descent Image Motion Estimation Subsystem (DIMES), it would prove a mission-saver, as events would show.

MER had started out as a simple upgraded rover adopting the same lander and EDL packages that had been developed and pioneered by Pathfinder. As the development of the dual mission progressed it became apparent that a mere upgrade was anything but realistic and a major test programme was essential on elements of the entry, descent and landing technology that had to be upgraded from the original concept. Pathfinder had demonstrated that the use of airbags was practical, and a lot on the previous mission had been taken from Viking technology and from the knowledge acquired during those two landings. But the rover itself was very different to Sojourner.

As an autonomous vehicle capable of controlling all events from shortly after launch through to the end of its mission across the surface of Mars, the two MER rovers had unique capabilities, and because of these it was necessary to carry out a wide range of tests before final assembly and launch. Having two rovers and not one helped rather than hindered the process. While it may be thought that two vehicles double the work, in reality it speeded up the process of qualifying systems, subsystems, components and science instruments. Because the two rovers were identical in terms of the engineering design, all these elements could be tested in parallel rather than in series, halving the time needed to clear the spacecraft for launch.

BELOW Inherited from Pathfinder, the airbag concept was modified and for MER would incorporate six-lobe bags for each of the four sides of the lander pyramid. *(JPL)*

ABOVE **Standing next to a full-scale model of a MER rover, Sofi Collis won a competition to name the two spacecraft, Spirit and Opportunity being chosen from among 10,000 entries.** *(JPL)*

At JPL the two spacecraft came together alongside each other, with sequential tests so that most equipment validated for one could be cleared for both. Some equipment, however, particularly the science instruments, had to be tested individually for each rover. Inevitably, although they were designed as identical twins each had its own peculiarities and each had a designation: MER-1 and MER-2. The first off the launch pad at the Kennedy Space Center would be Mission A, the second known as Mission B. Each spacecraft would get a name, selected after a competition launched by NASA in 2003 as a public relations exercise in association with LEGO, the toy construction company, which has always been highly supportive of space exploration.

More than 10,000 people submitted names for the two rovers. On 8 June 2003, the day of the planned launch, NASA Administrator Sean O'Keefe announced the winning entry. It was from a nine-year-old girl, Sofi Collis, of Scottsdale, Arizona. Sofi had been born in Siberia but was brought to the United States at the age of two and adopted by Laura Collis. She wrote: 'I used to live in an orphanage. It was dark and cold and lonely. At night, I looked up at the sparkly sky and felt better. I dreamed I could fly there. In America, I can make all my dreams come true. Thank you for the *Spirit* and the *Opportunity*'. And those were the two names selected. Because it was ready first, MER-2 would be named Spirit and fly Mission A, followed shortly by MER-1, named Opportunity, flying Mission B.

OPPOSITE **Spirit and Opportunity would be launched by Delta II rockets, with the initial stage erected first on Launch Complex 17A at the Cape Canaveral Air Force Station, Florida.** *(NASA)*

Planning the dual mission

The Cape Canaveral launch window for the MER flights opened at 2:06pm (local time) on 8 June 2003 and lasted for 16 days. Each spacecraft combination of cruise stage, aeroshell and lander would weigh 2,341lb (1,062kg) and include 115lb (52kg) of propellant. Within JPL the two rovers had been given designations as MER-1 and MER-2, named Opportunity and Spirit respectively. Whichever was ready first would launch as Mission A (also known as Rover A), the second – launched as early as 12:38 pm on 25 June – as Mission B, in a window lasting 20 days and closing on 15 July.

Delta II 7945 rockets would be used for both missions but the nine strap-on boosters for Mission B had about 25% more thrust than those for Mission A. The separation in launch dates was demanded by minimum turnaround times for changing ground equipment necessary for each flight. The first launch would take place from Launch Complex 17A, the second from pad 17B. The difference in launch times was driven by the relative position of the Earth on its 24-hour rotation, the more powerful rocket being required for the later launch date. Each spacecraft would fly a relatively fast Type I trajectory.

The duration of the cruise phase would depend upon the launch day, the flight path for each mission adjusted so that it would compensate for any delay in getting off the pad for weather or technical problems. Because of this, the flight times could vary from 194 to 210 days for Mission A and 194 to 214 days for Mission B. Engineers built in three trajectory correction manoeuvres up to the start of the approach phase, which begins 45 days before entry into the Martian atmosphere, when three further fine tweaks in the flight path could be made within the final eight days.

Early TCM burns would place the spacecraft on its correct course and compensate for any errors incurred by the launch vehicle, while the later manoeuvres would be made on the basis of a sophisticated technique pioneered by Mars Odyssey. On that mission, two traditional tracking modes had been reconciled into a hybrid computation for greater accuracy: ranging, which measured the distance between

LEFT The second stage is attached to the top of the Delta first stage inside a weather shield on LC-17. *(KSC)*

OPPOSITE The mated cruise stage and aeroshell are positioned on top of the second stage with a payload shroud encasing the assembled configuration to protect it during the dash through the atmosphere and into space. *(KSC)*

the Earth-based transmitter and the spacecraft; and Doppler, using the spacecraft's relative speed with respect to the Earth as defined by the frequency shift in the radio signal.

The reconciled calculation was then compared with a completely new method: delta differential one-way measurement. This combined two separate DSN stations on two different continents simultaneously receiving radio waves from the spacecraft with a distant radio source, such as a quasar, measured by the same stations. This triangulation technique was shown to improve the accuracy of the trajectory by several miles. The two MER spacecraft would use this to ensure best possible accuracy entering the atmosphere at precisely the right place to reach designated landing sites.

Now more than 40 years old, the Deep Space Network and its three primary sites at Goldstone (California), Madrid (Spain) and Canberra (Australia) were now each equipped with a single 230ft (70m) and at least two 112ft (34m) antennas, but the period from mid-2003 would place exceptional demands on the global system. Not only would they have to track and communicate with the two MER missions, but the European Space Agency was also launching its orbiting Mars Express spacecraft in the same window. In addition the DSN would be busy with NASA's Stardust spacecraft, which was on target to navigate the gas-filled coma of the comet Wild 2 on 2 January 2004 and collect dust samples, which would be returned to Earth in a capsule in 2006. NASA's Cassini mission to Saturn would also be nearing its destination, with great demands for trajectory measurements and commands for its orbital insertion anticipated for 1 July 2004.

During the early part of the cruise phase,

communication would be via a low-gain antenna on the cruise stage, which had the advantage that it does not need to be pointed continuously at the Earth, unlike later stages in the cruise phase when communications switch to a medium-gain antenna where the relative angle between the Sun (essential for solar energy) and the Earth becomes much less. During the entry, descent and landing (EDL) phase, communications would switch again to the low-gain antenna with a transmission rate of only 10bps, less than 2% that of the medium-gain antenna.

This and other low-gain antennas on the backshell, the lander and the rover would transmit a series of tones to indicate completion of critical stages on the descent profile. In all, 36 tones would be sent, each lasting ten seconds. After the lander emerged from the backshell an antenna would transmit a signal to Mars Global Surveyor, which would be within line-of-sight, and continue to transmit data for successive bounces of the airbags. Only after landing would the rovers use their high-gain directional antennas to transmit data direct to Earth, at a rate of 11Kbps.

After travelling 311 million miles (500 million km) from Earth, Rover A was planned to land within Gusev Crater at around 2:00pm local Mars time, or 8:11pm at the Jet Propulsion Laboratory, on 4 January 2004. Earth would be 105.7 million miles (170.2 million km) distant when Rover A touched down, with sufficient daylight remaining for the lander to open up and for electrical power to recharge the batteries. But communications between Mars and Earth would take 9min 28sec, and the same time back to the landing site.

Gusev Crater had been selected because it appeared from orbital images and remote sensing to have been a place where water had gathered to form a lake, possibly a small sea, in the distant past when Mars had a denser atmosphere and water flowed across its surface. Named after the 19th-century Russian astronomer Matvei Gusev, it consists of an impact crater approximately 95 miles (150km) in diameter located about 15° south of the equator.

The site is interesting because it lies at the end of a meandering valley 550 miles (900km) in length that appears to have been the product

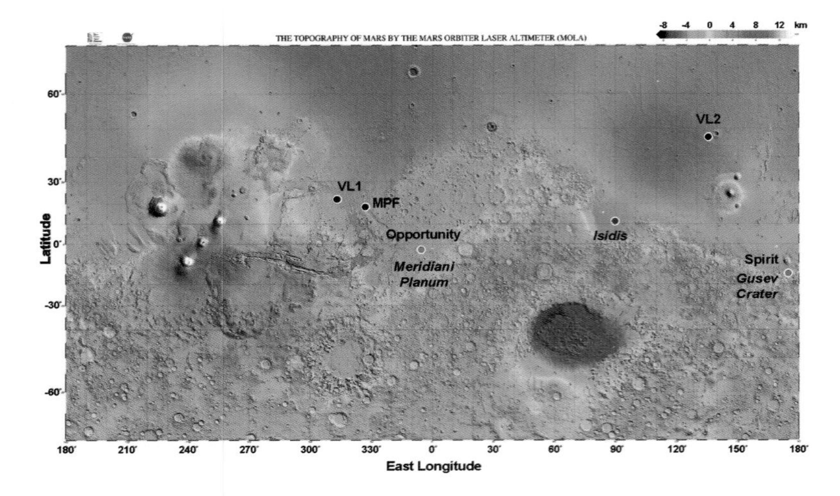

LEFT A topographic map of Mars shows that much of the northern hemisphere lies below the mean radius of the planet and appears to have been the region where oceans once covered much of the surface. *(JPL)*

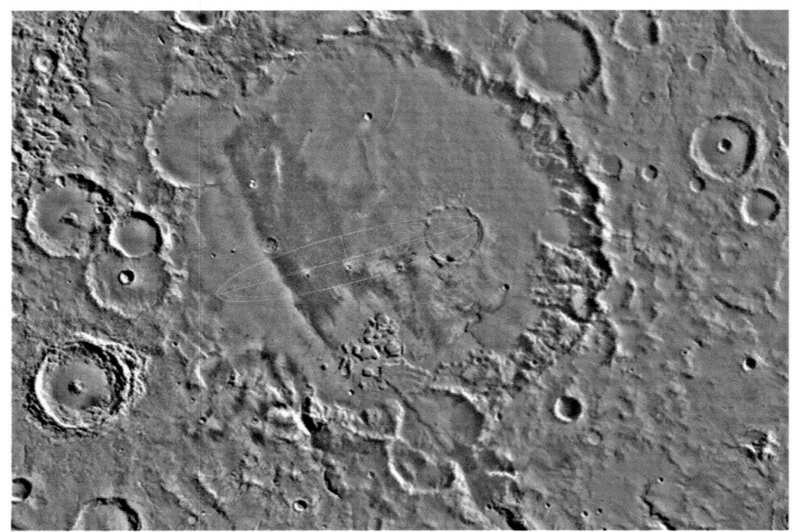

CENTRE Gusev Crater, the target for Spirit, lies at the end of a suspected water channel. The rover's entire life would be spent inside this crater searching for signs that Mars once had a habitable environment. Note the target landing ellipse. *(USGS)*

of a flowing water channel that breached the walls of this crater and flooded its interior, perhaps depositing sediments that could give clues to whether water really did flow on Mars. Answering questions about the potential that may have existed for primitive life forms was part of the Mars Exploration Rover mission and Gusev seemed well placed to provide clues.

The second landing would take place with Rover B touching down at a selected site in the Meridiani Planum at about 1:15pm Mars local time on 25 January 2004, 8:56pm at JPL, after a 305-million mile (491 million km) journey. Earth would be 123.5 million miles (198.7 million km) distant on that day. The site had been chosen because it too seemed to have had a watery past and some of the chemicals detected from orbit around Mars were likely to have been affected by water.

Meridiani is close to the Martian equator but around the other side of the planet to Gusev Crater. It is a region so named because it lies very close to the prime meridian as measured on maps. Observing the area from its vantage point in orbit, Mars Global Surveyor transmitted to Earth data from its thermal emission spectrometer to indicate that an iron oxide mineral called hematite is common in the area. The kind of hematite observed to be at the surface of Meridiani Planum could have been caused by water present in the distant past, but

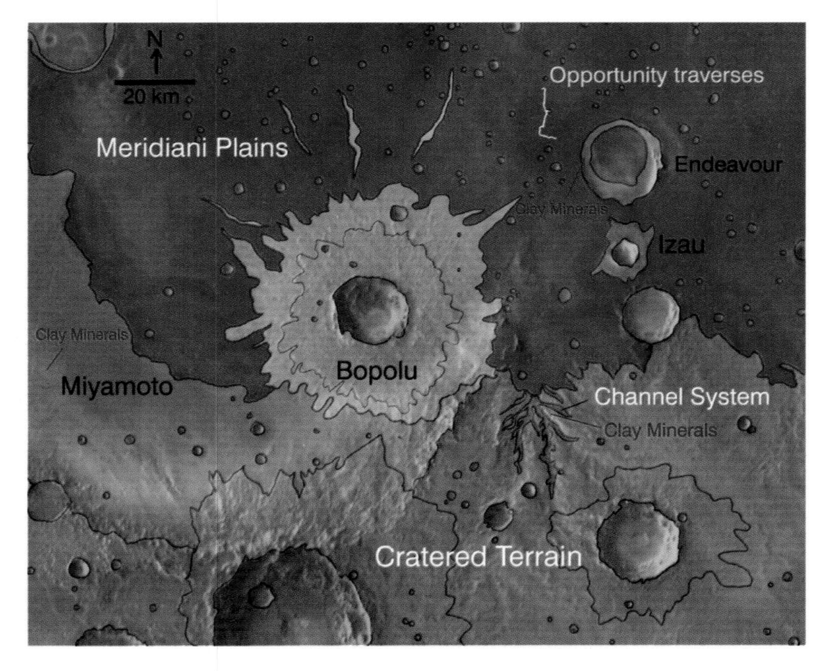

LEFT Meridiani Planum where Opportunity would explore lies close to the equator, and it too shows signs of having had its surface chemistry modified by flowing water. *(NASA)*

whether that was caused by water percolating through rocks or by trace deposits only on the surface of the rocks themselves was there for MER to discover.

The choice of landing sites had been large, scientists with specific instruments and conducting a particular science needed unique features, characteristics and surface conditions for their research. Initially 155 sites had been selected, but the choice had not been for the geology alone. Other considerations concerned engineering requirements. Apart from a wide range of criteria governing the EDL phase, including solar energy per time of day, probability of high winds aloft influencing the descent path, and the 'stiffness' (density) of the atmosphere, there were several surface conditions that had to be factored in as constraints.

The rock abundance of 20% for objects larger than 1.6ft (0.5m) had been carried over from the Pathfinder days but engineers felt that was too generous and tightened the number to increase the probability of success. And there should be no slopes greater than 2° over a distance of 3,280ft (1km) so that the kinetic energy released on bounce-out following touchdown could disperse, thus ensuring an energy-dissipating roll to a stop.

Cruise stage

With a weight at launch of 518lb (235kg), the cruise stage was the life-support structure for the lander and the rover during the seven-month trip to Mars. Combined, the cruise stage and the aeroshell had a diameter of 8.7ft (2.65m) and a height of 5.2ft (1.6m), the cruise stage comprising a disc-shaped structure with solar panels on the exterior and antennas on one side facing Earth. The cruise stage was fabricated from aluminium with an outer ring of ribs, a box section providing the structure between the outer and inner rings that carried the launch vehicle adapter.

The solar array was divided into five sections, three of which could be switched on or off via ground command as required. The panels comprised triple junction GaInP/GaAs/Ge (Gallium Indium Phosphorous/Gallium Arsenide/Germanium) arrays attached to an annulus on the cruise stage. They supplied a peak power level of 600W near Earth and 300W at the distance of Mars. The power bus was regulated by the cruise shunt limiter, with a radiator mounted to the inside of the launch vehicle adapter.

The telecommunications system included two X-band antennas (low- and high-gain) pointing in the –Z axis aligned with the spacecraft spin

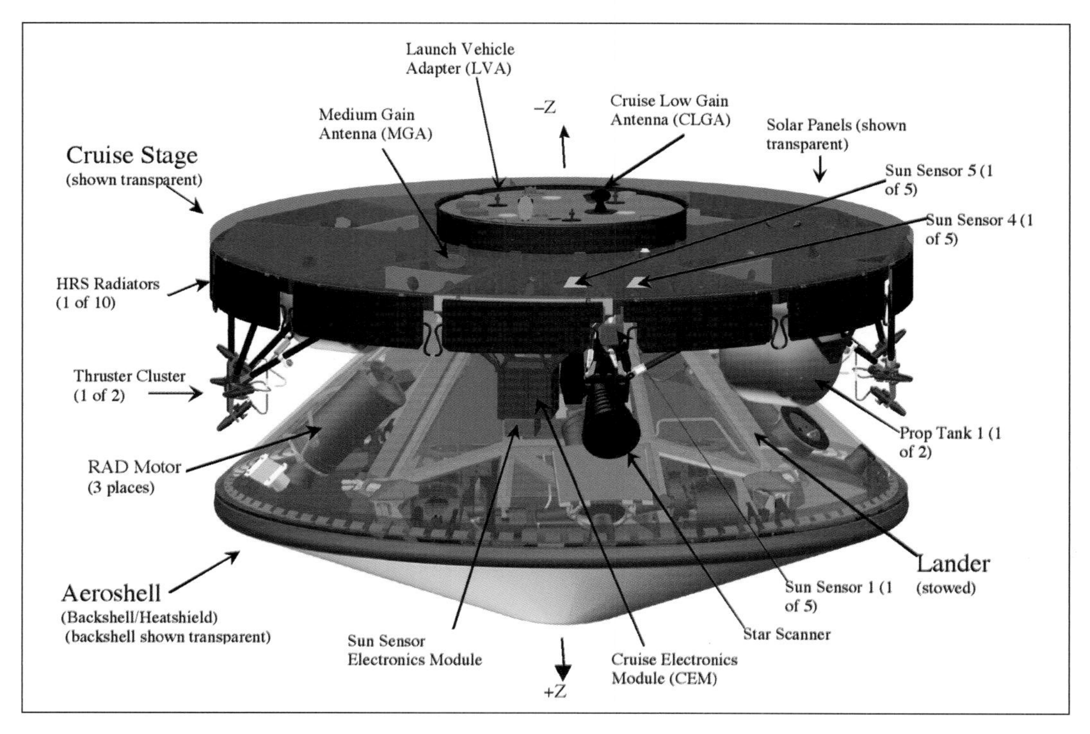

LEFT The cruise stage provided electrical power, communications with Earth, spin-stabilisation and attitude control and transmission of data from the spacecraft back to Earth, until separated prior to entry at Mars, where it would burn up. *(David Baker)*

RIGHT Build-up of Opportunity's cruise stage during May 2003 at NASA's Kennedy Space Center, prior to launch at Cape Canaveral Air Force Station. *(NASA)*

FAR RIGHT Propellant tanks and small rocket thrusters are clad in thermal insulating blankets to provide passive thermal control during the long flight to Mars. *(KSC)*

axis and directly on the stage-Sun line. The X-band system incorporated the Small Deep Space Transponder (SDST) supporting the two-way Doppler and ranging, command signal demodulation and detection, telemetry coding and modulation and 19MHz differential ranging (DOR) tone generation. The cruise electronics module provided an interface from the flight computer in the rover via a MIL-STD 1553 bus to the attitude and control system sensors and the cruise telemetry system.

Two diametrically opposed clusters of four 4.4N thrusters at 180° intervals were located on the rim pointing in opposite directions, each thruster mounted 40° off the X-axis (longitudinal). These were for spin control, attitude control and trajectory correction manoeuvres. There was also a star scanner and a Sun sensor that would be used for attitude determination and the navigation system providing data for calculating trajectory correction manoeuvres.

BELOW Almost ready for mating to the aeroshell, the cruise stage displays a mission logo at its centre. *(KSC)*

Two ultra-light titanium composite overwrapped propellant tanks situated inside the cruise stage each carried a maximum 68lb (31kg) of hydrazine monopropellant for the thrusters. Developed by engineers at JPL and a commercial company, the composite overwrap was a great asset to the weight-saving rationale. The cruise stage also had a Freon coolant loop for cooling the internal equipment and radiators situated on the outer rim.

The MER rovers

Compared to Sojourner, each MER vehicle was big. Deployed to the surface, it had a length of 5.2ft (1.6m) and weighed 384lb (174kg). At the height of the deployed solar panels, the rover had a width of 5.9ft (1.8m) and a length of 5.6ft (1.7m). It had a wheelbase 4.6ft (1.4m) long and 3.9ft (1.2m) wide, with a total height of 4.9ft (1.5m) to the top of the Pancam Mast Assembly. The core structure was made from composite honeycomb material insulated with solid-silica aerogel, and this formed a sealed Warm Electronics Box (WEB).

Aerogel is a remarkable material and consists of 99.8% air. It is 1,000 times less dense than glass and is extremely lightweight. Also known as 'solid smoke', aerogel prevents internal heat from escaping and is also a useful insulator against cosmic rays. All equipment sensitive to low temperatures was placed within the WEB, including avionics, flight computer, inertial measurement unit (IMU), rechargeable batteries and telecommunications hardware.

The WEB had a length of 34in (86cm), a width at the rear of 21.6in (55cm) and a depth of approximately 14.4in (36.6cm). The front of the WEB had a reduced width to conform to the triangular shape of the rover body and was designed to withstand a landing shock of up to 41g, with a working design load of 10g. The sidewalls were made from 5056 aluminium honeycomb panels bonded with an Astroquartz softening layer.

The top of the WEB formed the Rover Equipment Deck (RED), the top of which provided a secure mount for three antennas – the Camera Mast Assembly (CMA), the High-Gain Antenna (HGA) and a fixed panel of solar cells. Additional solar panels were attached by hinges to the sides of the triangular structure. These folded inside the lander while it was encapsulated within the three enfolding panels that would form the platform from which the lander would drive on to the surface of Mars.

Power for the rover came from 32 strings

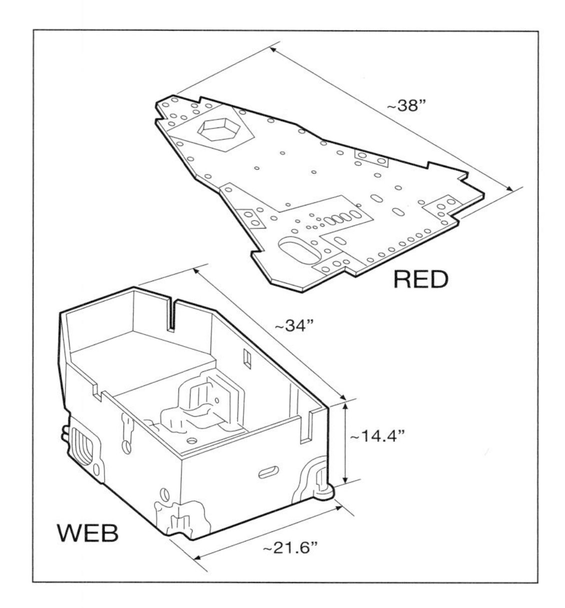

LEFT All electrical and electronics equipment, including the computer and sensitive processing equipment for the science instruments, went inside the Warm Electronics Box, where it could be insulated from low temperatures. *(KSC)*

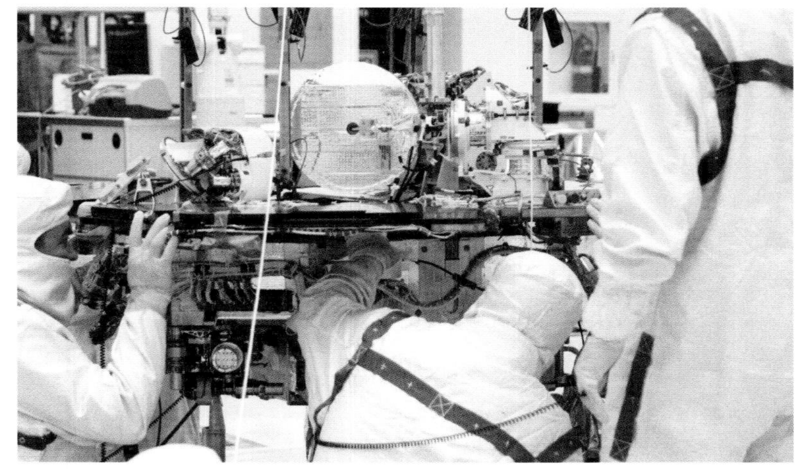

ABOVE The Rover Equipment Deck (RED) on top of the WEB carries structural supports for the High-Gain Antenna and the camera mast. *(KSC)*

LEFT The High-Gain Antenna (HGA) provided data flow between Earth and the rover and was mounted folded until deployment on top of the RED. *(KSC)*

RIGHT Electrical cables and wiring harnesses were attached snug against the main body of the rover but were left uncovered as part of the thermal control system and as a way of cutting down unnecessary weight. *(KSC)*

BELOW Opportunity shows off its delicate solar arrays system, with 32 strings of 16 cells each providing electrical power on the surface of Mars. *(KSC)*

BELOW RIGHT Each rover was checked thoroughly at the Kennedy Space Center after their trek across country from California to Florida, electronic diagnostic gear as much in evidence as it is in the servicing garage of a modern car! *(NASA)*

of solar cells on several panels, each string comprising 16 cells. The total area of solar cells was 14ft^2 (1.3m^2) of triple-layer photovoltaic cells, the same combination of Gallium Indium Phosphorous/Gallium Arsenide/Germanium cells adopted for the cruise stage. The cell panels could provide up to 900W/hr per sol at the beginning of the mission, predicted to reduce to 600W/hr at the end of the 90-day primary phase when dust had settled on their glass-like surfaces and the season on Mars entered a winter phase with low Sun angles.

Two 8amp/hr lithium-ion batteries recharged by the solar arrays were installed within the Warm Electronics Box since they had to be kept at temperatures above freezing, 32°F (0°C). The power system provided a nominal 28V DC, with the combination of solar-cell panels and batteries allowing the rover to draw more than 140W of peak power. The power bus was controlled by a two-stage shunt limiter with radiators integral with the deployable solar array panels. Each battery was controlled by a separate battery circuit board that was always

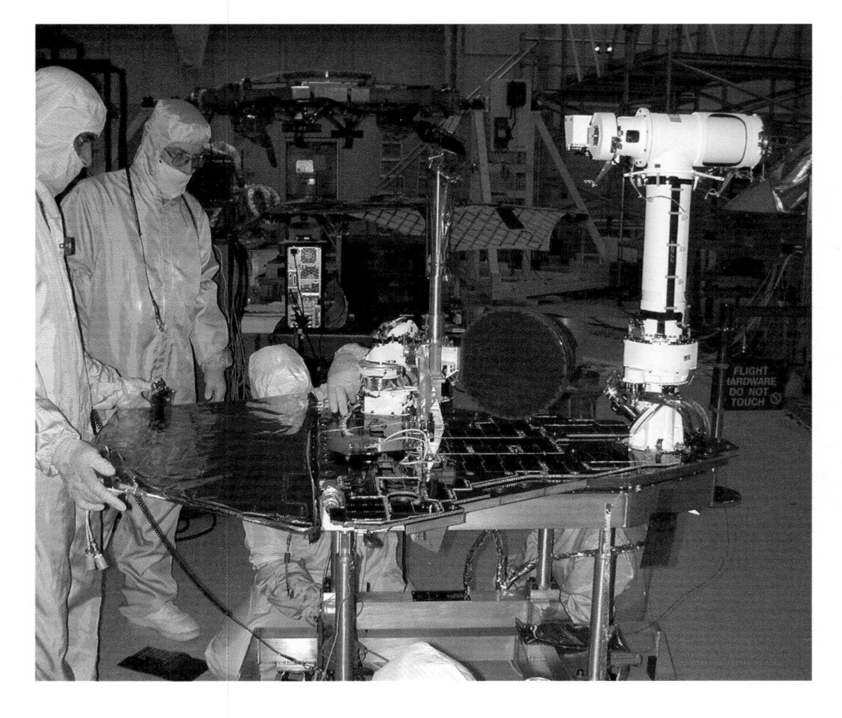

active, which automatically controls charging and discharging as well as cell balancing. This control board also gives 'wake-up' signals to the rover computer via a timer or when it senses adequate solar power flowing through the cell strings.

Thermal energy given off from electronics components, small electrical heaters and eight radioisotope units ensured that temperatures would never fall to freezing point. The radioisotope heater units each consisted of 0.1oz (2.7g) pellets of plutonium dioxide encased in a cladding of platinum-rhodium alloy wrapped by several layers of carbon-graphite composite materials. The technique is standard for several NASA spacecraft and had been used on the Sojourner rover to keep electronic systems warm.

Around half of the communications between the rover and Earth would go via the High-Gain Antenna (HGA), an 11in (28cm) diameter antenna beaming direct to the Deep Space Network stations over the X-band (8–12GHz) at 1.85Kbps. Because it is not omnidirectional, a two-axis HGA gimbal drives it to Earth lock. Identical azimuth (left and right) and elevation (up and down) motion went through a hemispherical cone of motion. Each used a 34V DC REO 20 Maxon brush motor spinning an integral three-stage planetary gearbox.

Mobility system

When it came to designing the wheel arrangement the proven experience with rocker-bogie systems, and the traction system for Sojourner, made this the logical choice for MER. Numerous rover test models by engineers at JPL and other NASA centres had advanced the state of the art, and by the time MER came along there was a valuable database on nuanced engineering tweaks that produced designs that were a considerable improvement on the early concepts.

The baseline requirement for the MER design team was twofold: provide a kinematic range that would allow the wheels to negotiate obstacles 10in (26cm) in height; and allow for 45° stability while absorbing a high percentage of the impact loads experienced during motion. In practice, the onboard computer would stop motion if the slope exceeded 30°. Moreover, while accommodating these primary requirements, the system had to be able to fold for stowage and unfold in such a way that would leave it optimised in the un-stowed stand-up configuration. Also, because the system had front- and rear-wheel steering it was free to do gradual turns or 'arc-turns', circular rotation on the spot.

As with Sojourner, where some scary

LEFT Tested by Sojourner in 1997, the two MER rovers made full advantage of a fully developed rocker-bogie design, allowing the vehicle to maintain traction on severe and undulating surfaces. *(JPL)*

obstacles were negotiated and clambered over, MER would utilise these capabilities to evenly distribute the load across all six wheels – what engineers call load equilibrium. This had the advantage of preventing all the wheels sinking into soft sand where only one wheel was affected, and because each wheel was integrated with its own drive actuator, each had its own differential. As with Sojourner, this reinforced the argument for making all off-road vehicle six-wheel designs with rocker-bogie suspension!

The need to get the rover off the lander could have required it to drop from a height that could cause damage to delicate instruments or science equipment on board. The suspension had to be a balance between a soft design that could place too much deflection on the rover body and one that could absorb shocks without transmitting those forces to the science equipment. The impact shock limit was set at 6g for any given force likely to be imposed on the rover during normal driving. And it all had to be encapsulated without compromising the deployment sequence or interfering with rigid elements of the rover structure.

Because both volume and mass were at a

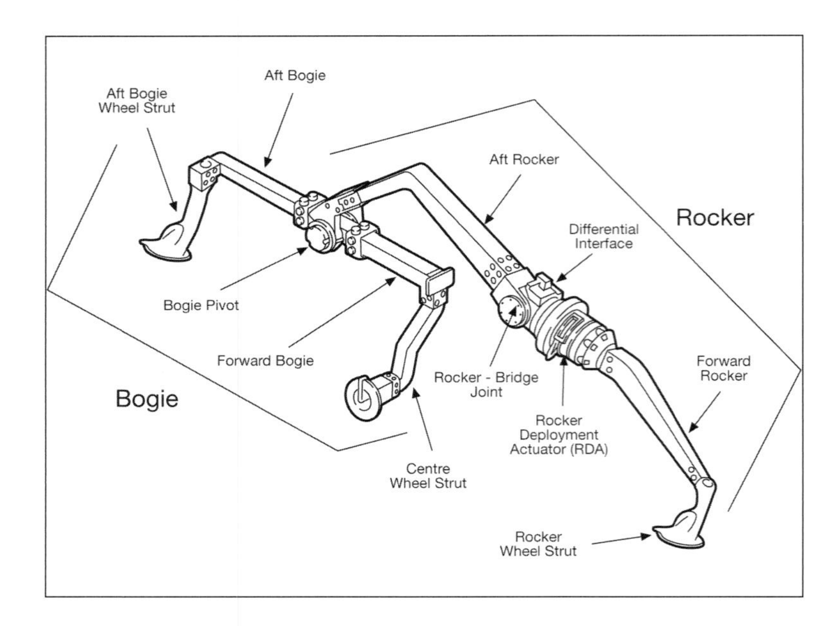

ABOVE The suspension for Spirit and Opportunity had several novel features to improve mobility, reduce friction, prevent lockup and ensure traction. Each wheel was equipped with a separate motor. *(KSC)*

BELOW Sojourner (front) had a fixed wheel geometry, but Spirit and Opportunity had a novel system for bringing the two front wheels together by rotating the front rocker wheel strut by means of an actuator close to the differential interface (see previous drawing). *(JPL)*

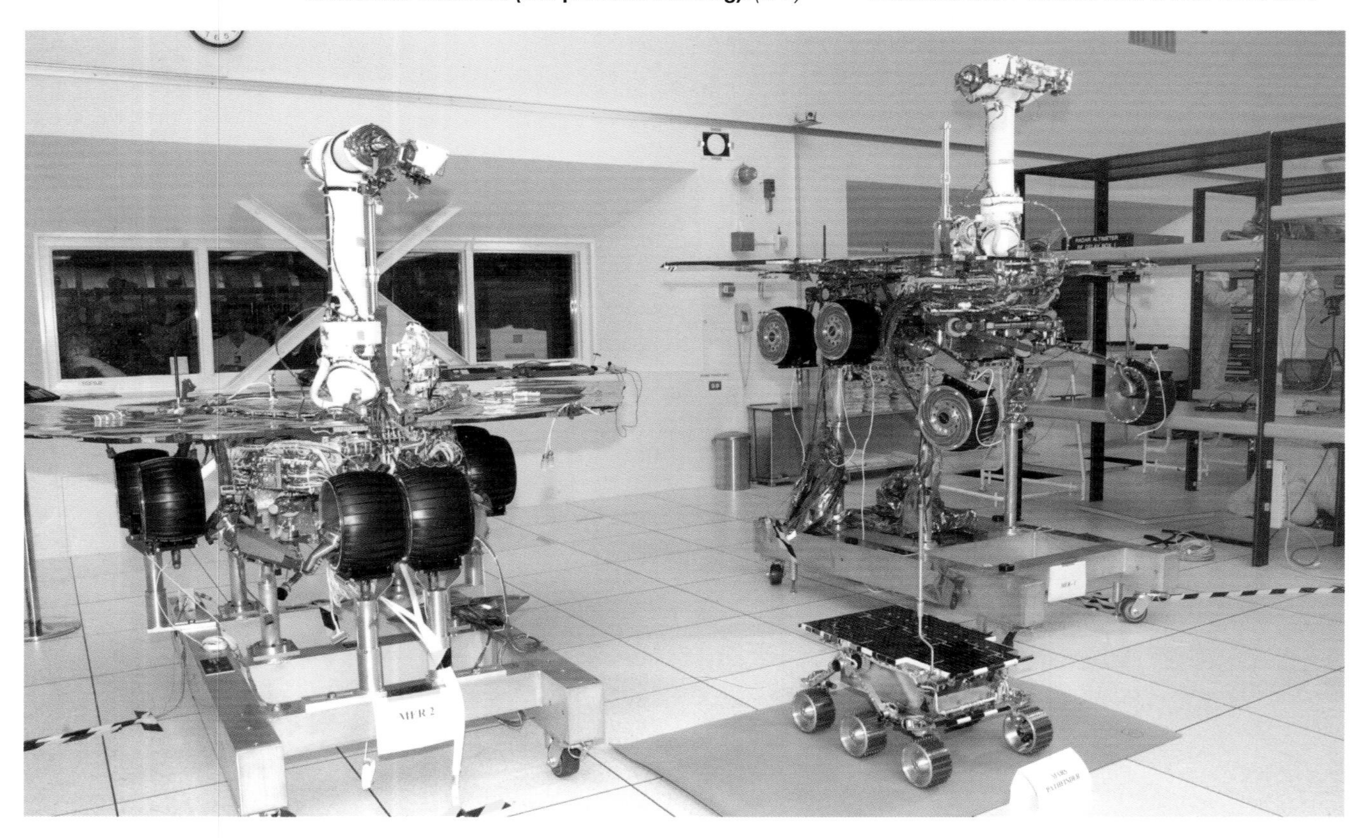

premium, titanium Ti-6AL-4V was selected for the structural components of the suspension, which had the added advantage of being welded, so optimising strength and flexibility. Eight of the ten suspension tube members were welded, left in the annealed state rather than being solution-treated, because the strength (900MPa, or 130ksi) was sufficient for all conditions. In fact, solution-treated state conditioning could have severely distorted the thin-wall construction of the tubes. Rover engineers needed to create a suspension that absorbed energy loads, and to maximise this the concept was one of a thin-walled box beam, with each beam tapered to save weight. The box beam design was good for both bending and torsional loads.

The method of joining was electron beam welding, unusual for space vehicles where proof loading is difficult to apply on the flight vehicle, which is why welding is usually limited to rocket propellant tanks, where pressure testing will readily disclose an imperfect join. Engineers had wanted to use investment casting but the cost for such a small number of parts was prohibitive. Required to withstand a maximum 6,325lb-in (714N-m) bending load should the centre wheel fall into a 7.8in (20cm) hole, the rocker-bridge joint adopted a yoke and clevis design, each one pivoting on two 52mm Torlon 7130 thrust/radial bushings with Braylon 601EF grease applied to ensure friction reduction. This joint could also survive a 4,475lb-in (506N-m) torsional load such as might be imparted from a lateral offset of 4.7in (12cm) between the joint itself and the wheels.

Stowage posed several problems. The rover was depressed and folded on the lander's triangular-shaped graphite-epoxy baseplate with its front wheels paired side-by-side directly in front of the main body of the rover. A rocker deployment actuator (RDA) was added to the suspension, a 4in (10cm) long, 3.1in (8cm) diameter, cylindrical unit with a Maxon REO20 DC brush motor with integral five-stage planetary gear head giving it a 175lb-in (20N-m) torque. The latch design was adopted from the solar panel latch on NASA's Magellan spacecraft. As the design of the rover progressed, the weight increased and the centre of gravity (cg) moved up.

Each wheel was made of aluminium and had

BELOW The inner faces of the two front wheels visible in the stowed configuration. Before driving off the lander the two forward rocker arms will rotate through 180° to place them in the same track as the rear pair. *(KSC)*

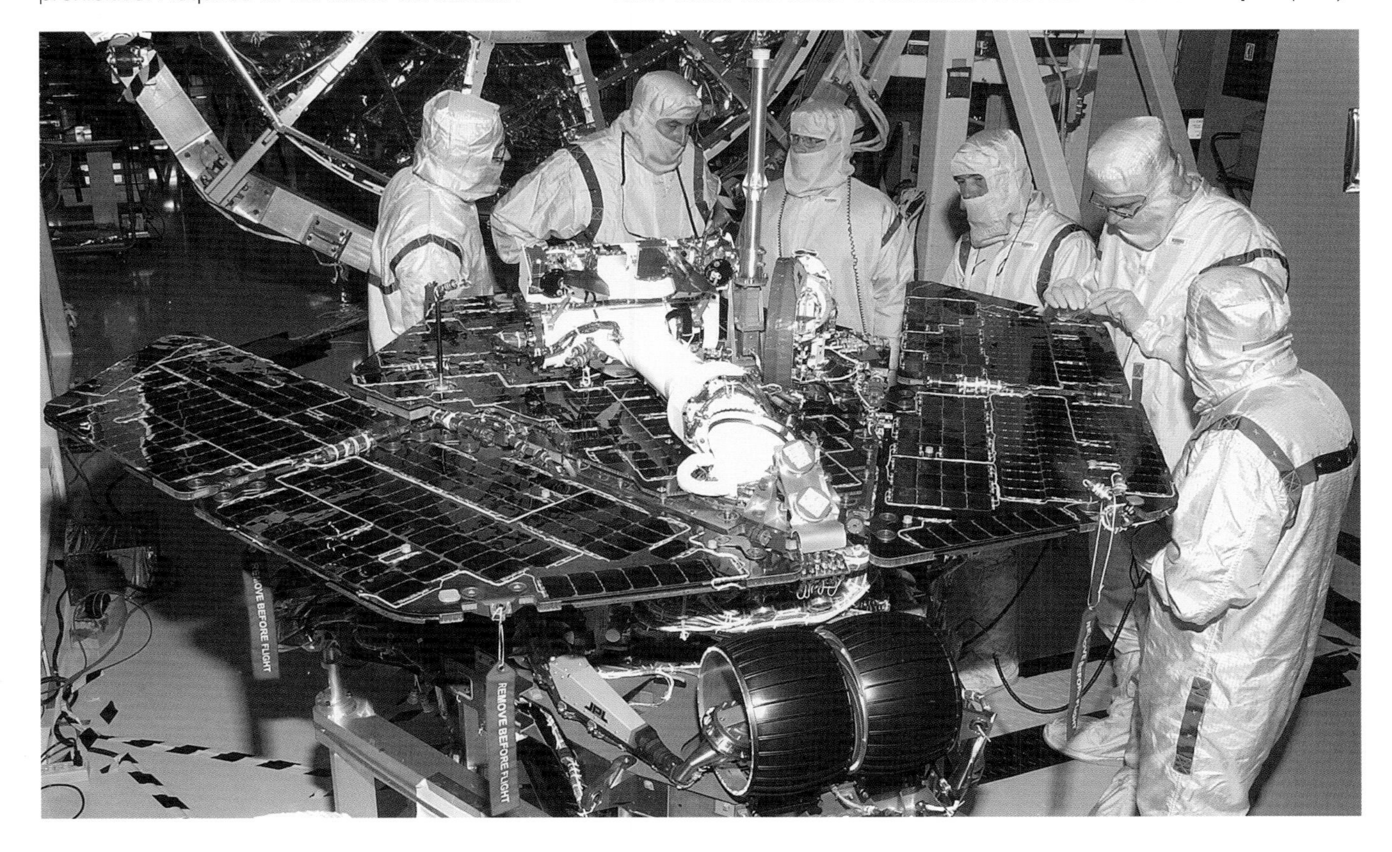

a diameter of a little over 10in (26cm), with spiral flexures to provide additional shock absorption, much like the additional cushioning afforded by pneumatic car tyres. The flexures were filled in with orange Solimide to prevent them becoming blocked with dust accumulations and rocky fragments. The offset displacement of the wheels on the rocker-bogie gave a length of 35in (89cm) between the centre of the front pair and centre of the middle pair, with 25.5in (65cm) between the middle pair and the rear pair of wheels. Each wheel had a width of 9in (23cm). The track of the three wheels on each side was 57in (145cm), equally spaced 28.5in (72.5cm) either side of the rover centreline. To honour the 45° tilt requirement, the length of the bogie had to be increased by 6.7in (17cm). On a hard surface the normal maximum driving speed was 0.16mph (0.26km/h), but about 0.03mph (0.05km/h) under hazard avoidance autonomy.

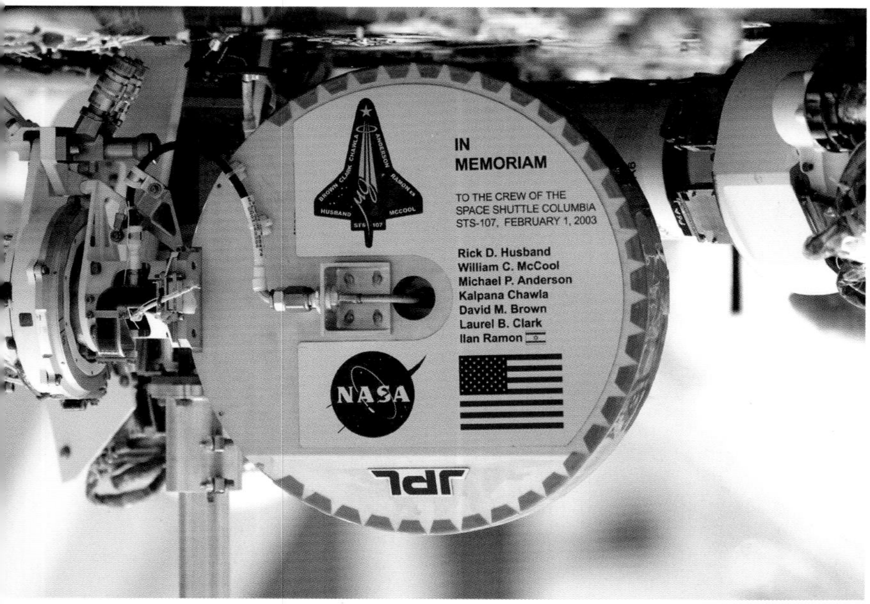

Autonomous control

A major departure from the Pathfinder design was that for Mars Exploration Rover missions all flight system command, data handling and motor control functions were located inside the rover using a VME bus and a 32-bit RAD 6000 micro-processor. This is a radiation-hardened version of the PowerPC which at the time was used in some Macintosh computers, operating at a speed of 20 million instructions per second. The 128MB of RAM was augmented by 256MB of non-volatile flash memory that allowed it to retain memory without power.

Due to the greater abundance of electronic and electrical equipment inside the rover, this brought special requirements for cooling. As with the basic method adopted for Sojourner, each MER vehicle would encase its electronics in the WEB, but the 125W generated during the cruise phase required an active Heat Rejection System (HRS). The main feature of the HRS was two redundant pumps for circulating the CFC-12 cooling fluid, but with an accumulator to accommodate changes in the large variations in cooling fluid circulated at different times.

Maintaining a safe temperature within the WEB was an essential feature of keeping the rovers alive, and they would achieve that through a combination of thermostats and heaters. The batteries were particularly sensitive to temperature and yet they were arguably the most important part of the 'life'-support system for each rover. The batteries were packed in the WEB along with eight RHUs (Radioisotope Heater Units), each producing 1W of energy per day. Thermal switches would activate or deactivate the heaters if temperatures strayed outside the –4°/68°F (–20°/+20°C) band. Too low and the heaters would turn on; too high and the heat transfer path would open to increase thermal flow to the radiators, dispersing excess energy to the atmosphere of Mars.

The Rover Electronics Module (REM) contained the electronic boards and scabbed-on telecommunications instruments and this was situated within the WEB. The cruise stage had 10 radiator panels compared with 12 on Pathfinder, together with a Heat Pump Assembly (HPA) with associated motor control units, check valves, and thermal valves to bypass the radiator designed by Howden Fluid Systems and known as the Integral Pump Assembly (IPA). Shortly after the design was completed, Howden moved to a different location and the flight hardware was built by Pacific Design Technologies, with full sets being delivered to JPL by October 2002. The WEB on Mars Pathfinder had been capable of rejecting a heat load of 90–180W at a radiator temperature of –112°F (–80°C) to 68°F (20°C), and this was found to be adequate

BELOW The back of the High-Gain Antenna displays a commemoration to the seven astronauts whose lives were lost when the Shuttle *Columbia* was destroyed returning to Earth on 1 February 2003, only a few months before the launch of Spirit and Opportunity. *(KSC)*

RIGHT Opportunity gets plugged into an electronic test to evaluate systems, tests conducted on each rover now being an essential part of defining the behavioural signature of each vehicle. *(KSC)*

for MER. But the plumbing was significantly different. However, the total heat load was higher than for Pathfinder, and the thermal density was higher too.

The cooling system for Pathfinder had been relatively simple, since the electronics were all on a two-dimensional shelf. The Rover Electronics Module (REM) for MER was much more complex, and for ten months the design moved back and forth to eliminate a succession of unexpected problems. On Pathfinder the electronics shelf was made of aluminium with a thickness of 1.5mm, with the face sheet thickened to accommodate a solid state power amplifier. For MER, the REM was fabricated from Aluminium Al7050-T7451 and measured 10in x 11.4in x 17.6in (26cm x 29cm x 44.6cm), the HRS tubing being carried across the REM with a thickness of 0.0028in (0.71mm). Tube length was about 200in (5m), with 46 bends that increased the effective length by 40%. The cruise stage radiator for Pathfinder utilised 12 panels distributed evenly around the outer circumference with a capacity for rejecting a maximum 180W. It was manufactured from

ABOVE Assembly of the rovers begins with the lander base plate and its three associated petals. *(KSC)*

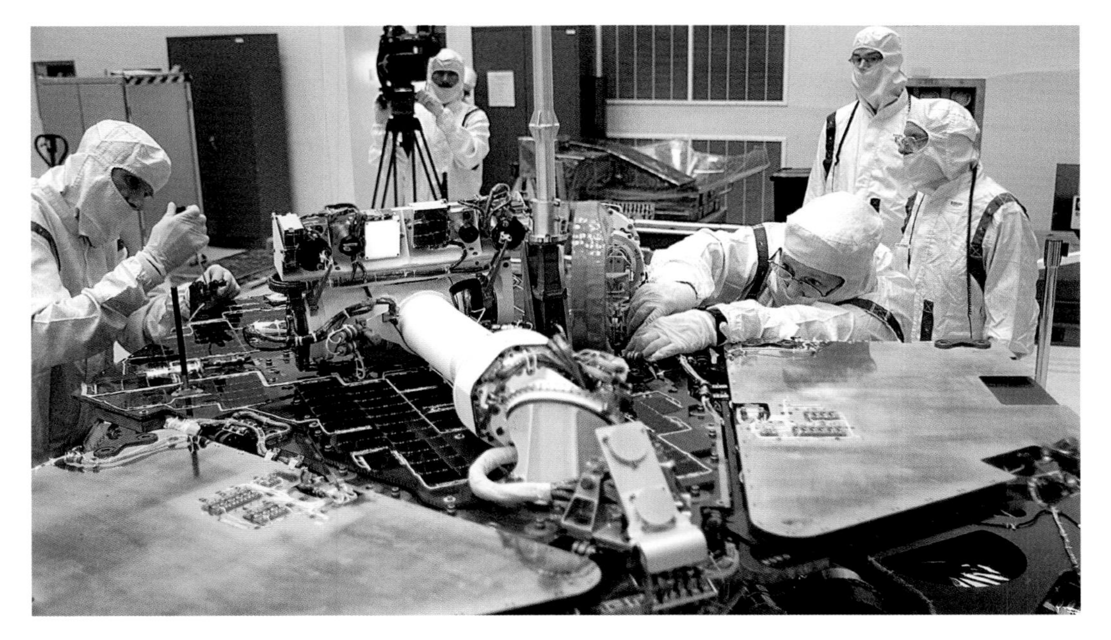

LEFT Opportunity is given a fit-check inside the lander. *(KSC)*

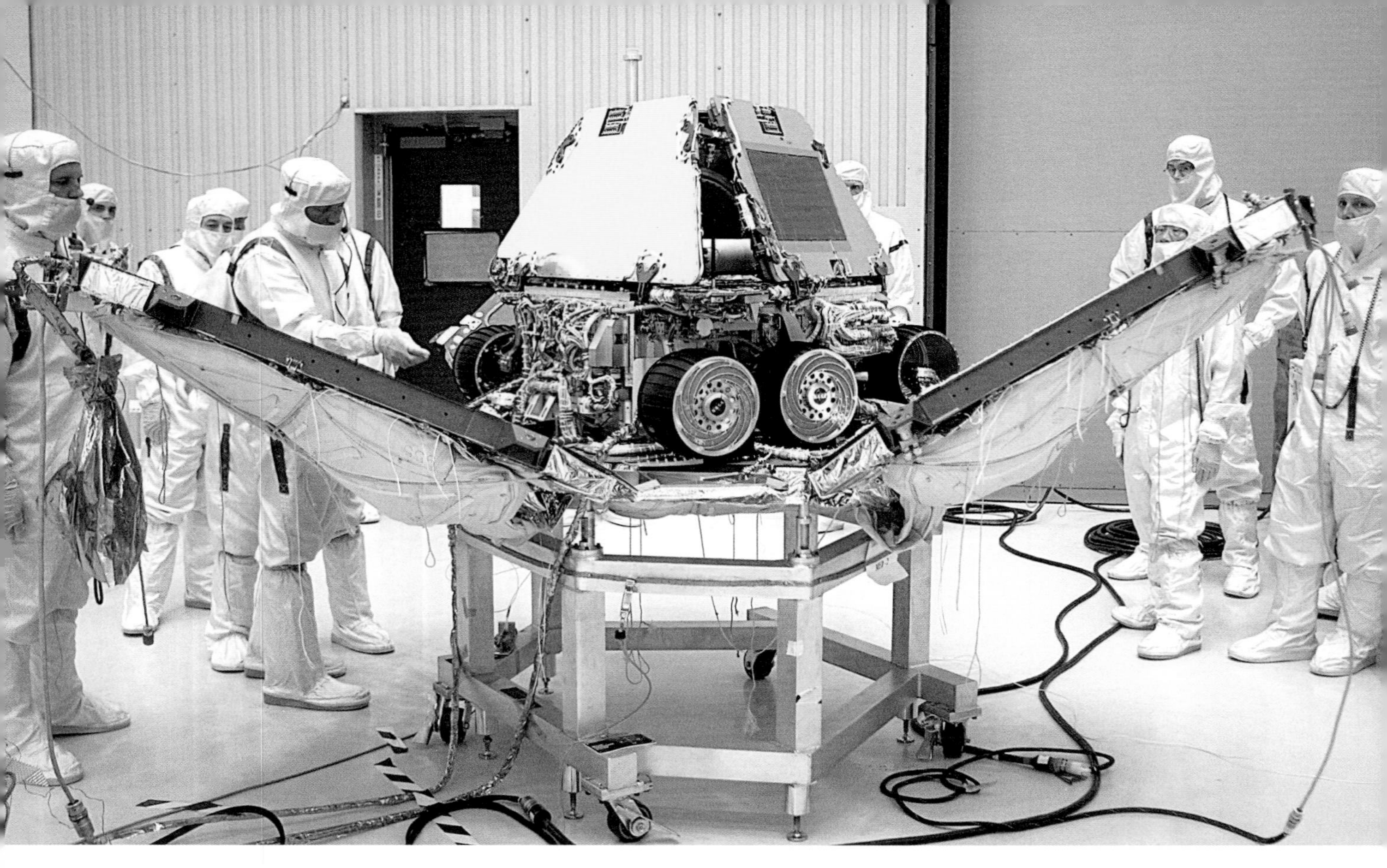

ABOVE From the inward folding front wheels to the reduced track for the middle wheels, and from the shape of the solar-cell arrays to the configuration of the rover deck, the shape of the rover was dictated by the confines of its lander shell. *(JPL)*

aluminium with a thickness of 0.75mm and thermally attached to 3/8in (9.53mm) diameter tubing also made of aluminium.

During its flight to Mars the rover was encapsulated within the lander, which was itself surrounded by airbags, the whole assembly inside an aeroshell manufactured by Lockheed Martin. Nevertheless, the REM was the heart of the control system for all flight activities from separation away from the launch vehicle through to the end of its mission on the surface of Mars. To do all this, the computer is a 32-bit Rad6000 processor with a 20MHz clock rated at 20MIPS using the VxWorks operating system. It has an onboard memory of 128MB of DRAM and 3MB of EERPROM. The non-volatile memory and camera interface board has an additional 8MB of EERPROM as well as 256MB of flash memory for storing telemetry data.

Software in the main computer provides a control loop monitoring the status of the vehicle, checking for commands to execute, performing communications and checking health and condition of the rover. It is responsible for checking temperatures, especially within the Warm Electronics Box, responding to potential temperature problems and formatting telemetry into data packets either for direct transmission to Earth or via an orbiter, or for receiving and decoding commands. The rover monitors itself continuously and assembles engineering information for periodic reports to Earth, either directly or stored for later transmission.

Entry, descent and landing

As the spacecraft neared Mars it would transition to the EDL phase using software known as 'EDL behaviour' approximately two hours before entry. Among other functions this called for the HRS to vent the Freon system just over one hour prior to entry. With the coolant fluid gone the temperatures would begin to rise proportional to the duty cycle of the equipment and the thermal capacity of the components. For this reason the temperature was chilled down with steady-state temperatures at the inlet to the REM being 5°F (–15°C) at entry. A significant problem occurs in the very process of venting the Freon because any free stream of particles will act like a thruster and upset the stability of the aeroshell. On Pathfinder this was predicted to cause a nutation (wobble or

ABOVE Actuators and electrical connections will be carried on and within the fabricated faceplates of the four petals, including the base plate. *(NASA)*

coning effect) of up to 0.25°. In fact it induced a nutation of 2.5°, which would be totally unacceptable for MER.

Because the EDL phase for MER had tighter constraints so as to reduce landing errors to a minimum, any nutation effect would be like a thruster shifting the aeroshell away from its desired flight path. If only by a tiny margin, this would be magnified lower down in altitude as a miss distance of several miles. A detailed study was conducted to solve the problem but nobody could come up with a definitive conclusion as to why the nutation on Pathfinder was so high and what to do to reduce it to the minimum for MER. To combat all the possible conditions that could have resulted in the tenfold increase in nutation, the diameter of the vent tube was increased so that expulsion of all the fluid would take 60sec instead of 2min 30sec, chamfering the exit plane of the nozzle to prevent back-impingement of the gases at MER's supersonic entry, and relocating the vent nozzle so that the propulsive effect of the vent would be virtually nil.

About 70 minutes before entry the thrusters would position the spacecraft to the correct separation attitude, which occurs 15 minutes before that event. Just 50 seconds before separation the onboard X-band system would start to send signal tones (Multiple Frequency Shift Key, or M-FSK, tones) coded to indicate the state of the spacecraft and to acknowledge critical events as they occur. After separation, communication with Earth would be via the low-gain antenna on the backshell.

The MER entry mass of 1,840lb (835kg) was considerably greater than the weight of Pathfinder at entry (1,290lb, or 585kg), with local time being in the mid-afternoon, resulting in a less dense atmosphere than Pathfinder found. Moreover, the landing site was at a higher altitude above the mean planetary radius, providing a less dense atmosphere on the elevated surface. This combination of extra mass and less dense atmosphere presented a higher terminal velocity due to reduced deceleration during entry and shorter period of time to arrest the speed of the spacecraft. The geometry of the entry trajectory and the phasing of events were essential to getting the heavier spacecraft on the surface.

The 12,960mph (20,865km/h) inertial velocity of MER at entry was about 20% less than that of Pathfinder at a flight path angle of −11.5°, compared to a steeper −14.2° for its predecessor. This was forced by a need to accommodate the additional mass while satisfying a 1° margin of error. The 198lb (90kg) heat shield for the MER was based on the Pathfinder shield with a 70° half-angle sphere cone manufactured by Lockheed Martin. The forebody was protected by a lightweight SLA-561 ablation material and the 49° backshell was covered by a spray-on equivalent. The backshell had a 0.65m orifice through which the parachute canister would deploy the parachute.

Consisting of an aluminium honeycomb structure sandwiched between graphite, epoxy face sheets, it was covered on the exterior by a phenolic honeycomb material. Made up from benzene, the honeycomb was filled with an ablative material comprising a unique blend of cork wood, binder and tiny silica glass spherules. The ablative material covered both the backshell and the heat shield, but the latter had a thicker 0.5in layer covered by a very thin aluminised blanket to protect it from low temperatures during the cruise portion. The backshell was painted white. The aeroshell also carried a Litton LN-200 inertial measurement unit used with the rover's IMU to measure deceleration and trigger deployment of the parachute. Together, the backshell and the parachute weighed 436lb (198kg).

Although the spacecraft was heavier, the slower entry velocity and shallower entry angle compared to Pathfinder produced a less severe aerothermal environment. This allowed for a reduction in ablator thickness from 0.75in (1.9cm) to 0.64in (1.63cm). Nevertheless, the peak temperature would reach 2,637°F (1,447°C) with peak heating at 50W/cm^2 versus 105W/cm^2 for Pathfinder. Total heat load would be 3,400J/mcm^2 compared to 3,900J/cm^2 for MER's predecessor, with a peak deceleration of 10g versus 20g for Pathfinder. The peak stagnation pressure for MER was 25 kN/m^2, which was only a little higher than the stagnation peak pressure for Pathfinder.

Because of the higher entry mass and the less dense atmosphere, the time between parachute deployment and retro-rocket ignition was much less compared to Pathfinder. To allow sufficient time for the necessary events to take place during descent, compared to Pathfinder the parachute deployment altitude was raised, the deployment algorithm was modified and the size of the parachute increased. The deployment system was triggered by a dynamic pressure setting of 725N/cm^2 compared to 600N/cm^2 for Pathfinder, which corresponded to an altitude of about 8.9km. In addition, the deploy algorithm was based upon dynamic pressure and not on a deceleration value as used for Pathfinder. The maximum dynamic pressure observed by MER was calculated as 800N/cm^2 rather than the 703N/cm^2 of Pathfinder.

The design of the parachute was heavily influenced by the increased mass and the deployment altitude. Well proven by Pathfinder, the Disc Gap Band concept was now to demonstrate it could be adapted for a more demanding requirement. The key aspects of its design gave the DGB flexibility. With a central hole at the top, the diameter of which had proven critical during testing for satisfactory operation by Pathfinder, the disc shape provided retardation while the gap ensured stability following inflation and helped minimise sideways motion. JPL's Adam Steltzner played a major role in crafting the design for MER and tests were carried out from an airdrop by helicopter to a workout in a giant wind tunnel at NASA's Ames Research Center. It was from this painstaking work that the final configuration of the basic design emerged, and not without pain en route!

The diameter of the parachute was increased to 49.5ft (15.09m), which increased its area by 22%. The effective drag area was increased by 28%, this being accomplished by a reduction of 5% in the band length. The parachute had a drag area of 721ft^2 (67m^2) compared to 565ft^2 (52.5m^2) for Pathfinder, and a trailing distance of 9.4 body diameters, identical to Pathfinder and thereby avoiding wake interference issues. Inflation pressure was 18,000–19,000lb (80,100–84,600N), compared to about 8,000lb (35,600N) for Pathfinder. The parachute itself was fabricated from polyester and nylon with a triple Kevlar bridle connecting it to the backshell. Incorporating 48 suspension lines attaching it to the backshell, the parachute was to be deployed approximately 4min 1sec after entry interface at an altitude of 5.3 miles (8.5km) and a velocity of 960mph (430m/sec).

About 20 seconds after parachute deployment, the heat shield would be jettisoned by pyrotechnically severing six nuts and releasing the energy in push-off springs. The lander would then rappel down a metal tape on a centrifugal braking system carried in one of the lander's four petals. The slow descent would bring it to the full extent of another tether 65ft (19.8m) long. Made of braided Zylon, the tether material was similar to Kevlar and provided space between the spacecraft and the backshell/parachute combination to deploy the airbags and maintain a safe distance from the retro-rocket exhaust plume when they ignited. The bridle incorporated an electrical power line to control the firing of the rockets and to carry information on rate and tilt from the backshell to the flight computer located inside the rover. The MER spacecraft would take a nominal 100sec to reach the surface after parachute deploy, but could be as short as 65sec.

The retro-rockets for MER had 70% greater impulse than those on Pathfinder and were known as the Rocket Assisted Deceleration subsystem (RAD), arranged in a symmetric cluster of three along the inside surface of the backshell. They would provide a total impulse of up to 95,000N-sec. At an altitude of 1.5 miles (2.4km) the radar altimeter would acquire the surface, and this data would be used by the flight

TOP Its aeroshell in the background, Opportunity is positioned on top of the lander base section. *(KSC)*

ABOVE The petal wings open revealing the 'bat-wings', designed to prevent the rover slipping on to its side should the lander come to rest on a tilt. *(KSC)*

LEFT The lander is offered up to the rover, the side curtains being tucked in between the side panels. *(NASA)*

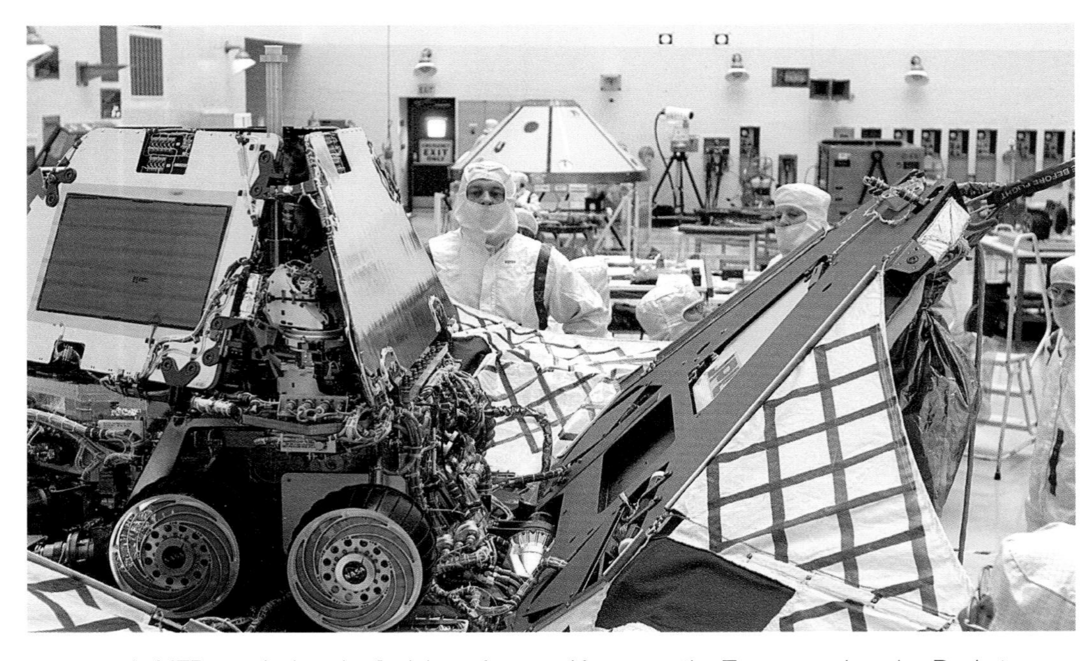

RIGHT The aft section of Opportunity displays the abundance of electrical wiring and cables connecting the rover to the lander, from which the lander will get commands to open the panels and draw in the deflated airbags. *(NASA)*

computer in MER to calculate the final times for airbag inflation followed by retro-rocket ignition 2sec later. The flight computer would aim to ignite the RAD at a height of 425ft (130m), the goal being to bring the package to zero descent rate 49ft (15m) above the ground and release the bridle. Relieved of the weight of the MER load, sufficient impulse would remain in the retro-rocket to carry the backshell and parachute to a safe distance away from the lander as it free-falled to touchdown and its first bounce.

To improve landing accuracy and to counter wind shear and high winds close to the surface, the MER spacecraft carried steering rockets added to the EDL systems. Known as the Transverse Impulse Rocket System (TIRS), they comprised three 2kN-sec impulse solid propellant rockets. Arranged at 120° intervals around the circumference of the backshell, they were canted outward 85° to the thrust vector of the RAD rockets, placing them almost at right angles to the vertical. After the radar had acquired the surface, a descent imager would take three pictures at 4sec intervals to determine the transverse, or sideways, movement of the vehicle as it descended, effectively determining if and to what degree the spacecraft was being blown off course. The computer would then calculate the force needed by the TIRS rockets and the combination required to counteract this force and stabilise the descent.

BELOW Airbag and curtain activators are carefully aligned in final closeout prior to installation inside the aeroshell. *(KSC)*

Tests with the EDL systems on Earth prior to launch revealed that in around 10% of cases high lateral velocities caused by winds could threaten the bags, stressing them far beyond their design limits, and the remedial efforts made to prevent that were vital. Wind shear could cause the aeroshell to oscillate, swinging from side to side by up to 20° so that when the RADs fired to slow the descent just 6sec before reaching the surface, the retro-rockets would be pointing at an angle and could contribute to the sideways motion, making it much more likely that the airbags would strike the surface while moving sideways, tearing the fabric and wrecking the mission before it started. The

LEFT Packed tight and ready for transport from Earth to Mars, the Opportunity rover is snug inside its protective arrangement of 24 airbags. *(NASA)*

TIRS would be insurance against the airbags swinging and ripping to pieces against rocks.

The landing cushion for MER comprised four separate interconnected airbag assemblies, one for each face of the tetrahedron shape formed by the folded lander. Each comprised six lobes 5.9ft (1.8m) in diameter, made from Vectran and held together so that impact loads from one could be shared by adjacent segments. The bags were not attached directly to the lander but restrained around it with ropes forming a net across the structure. With nominal inflation pressure of 1.0psi, when fully inflated, the airbag cluster had a total height of 14.9ft (4.56m), resting on a square 11.6ft (3.53m) on a side. Each lobe had a diameter of 4ft (1.14m). The airbags looked the same as those used on Pathfinder but they were very different. Each had a double bladder system whereas Pathfinder's airbags had a single bladder and four layers of scuff-resistant 100-denier Vectran, cut with 6% fullness and attached on top of the bladder.

Tests had shown that a system whereby the inner sealing membrane was cut with fullness relieved the gas membrane from induced stresses. In tests carried out at NASA's Glenn Research Center, Plum Brook Station facility, it demonstrated that with this design, the 1,212lb (550kg) MER package would be resistant to an impact of up to 85ft/sec (26m/sec), or 56mph. This was the same as for Pathfinder weighing only 683lb (310kg) at impact. Remarkably, and not a little due to changes and improvements to the parachute system, the same basic EDL architecture as Pathfinder carried 80% more payload, 981lb (445kg) in MER versus 551lb (250kg) for Pathfinder.

The lander airbags were to inflate about eight seconds prior to touchdown, followed two seconds later by the RAD retro-rockets and, if needed, the TIRS thrusters. Three seconds later the bridle would be cut on command from the main control computer and the airbags encapsulating the lander would be in free-fall to the surface. In those three seconds it would drop from a height of 33–49ft (10–15m) and bounce back up to a height of perhaps 49ft (15m) before recontacting the surface and repeating the cycle 15 or 16 times, rolling to a stop up to 0.6 mile (1km) from the initial impact spot.

The lander

The lander itself weighed 805lb (365kg) and comprised a primary structure consisting of four composite petals with titanium fittings, the base petal underneath providing a mounting platform for the three side petals, operated by high-torque actuators. These actuators could be independently adjusted from the stowed (tetrahedron) configuration to the 'iron-cross' configuration, and could deflect the side petals below the level of the base petal to adjust for

RIGHT **The 25ft Solar Thermal Vacuum Chamber at JPL can put a spacecraft through its entire mission cycle for radiant solar energy simulation and test its ability to control internal temperatures.** *(JPL)*

surface contours and thereby provide a flat platform across which the rover could drive to the surface. These adjustments could be made by ground command.

But getting the rover off the lander on to the surface required the deflated airbags to be drawn back in underneath the lander. Tests showed they could bunch up or form balls of cloth that would obstruct the free passage of the rover from the lander to the surface. The method used to deflate the bags and draw them underneath and closely around the petals of the lander structure used essentially the same technique as that for Pathfinder.

Four Airbag Retraction Actuators (ARAs) attached to the petals were driven by brush motors that were controlled by electronics in the rover. These caused the bags to deflate, allowing the lander to open. Three Lander Petal Actuators (LPAs) operated by brushless motors opened the side panels and allowed the lander to ease itself on to its base platform, where the rover was situated. Retraction of the airbags and petal deployment could take between 75 minutes and 166 minutes depending on which petal was base-down.

BELOW **The heat shield would not experience the spikes in temperature seen by Pathfinder, but now the technology was well proven and it would fly a more benign trajectory.** *(KSC)*

When the petals were fully open, silicone-coated Vectran cloth surfaces (dubbed 'bat-wings'), supported by ribs and cables connected directly to the lander petals, were to be deployed to link the spaces between the petals, giving them stability. Vectran is like Kevlar but has greater resistance to low temperatures and was the same material as the lander airbags. They were extracted automatically by the petals as they unfolded and this 'passive deployment' obviated the need for pyrotechnic devices to deploy them. Both petal covers and bat-wing panels were criss-crossed with black lines over the light coloured fabric, so that ground controllers could see if they were deployed properly. In this form they make an approximately circular platform that can be lowered flush with the surface, thus avoiding a large step down to the ground.

The precise sequence of operation would depend upon the condition at the surface and on whether there were any obstacles impeding the deployment of the lander. The nominal sequence would begin 12 minutes after touchdown (L+12) with airbag retraction, followed 15 minutes later by firing the latches holding the petals in the folded position. Full airbag retraction would be finished at L+67, followed by the petals deploying two minutes later. Under normal base-down conditions, the lander would deploy at L+89, when the petals would be fully open. Solar array latches would be fired with the primary array deploying immediately thereafter, followed by secondary array deployment six minutes later.

With primary and secondary solar panels fully open, the rover would switch to its secondary batteries and the system would place itself in a 'safe' condition taking energy from the Sun. Still under the 'EDL behaviour' phase, the rover would deploy its Pancam Mast Assembly (PMA) and the High-Gain Antenna (HGA), allowing the rover to take a panoramic sweep of the surrounding area to tell flight controllers what sort of conditions the rover found itself in and whether the airbags had retracted fully to give the rover a roll-off path to the surface free of obstructions.

The Pancam images would provide the broader view, but the Hazcam images allowed controllers to determine the step-off height for the rover. The Navcam pictures provided visual

ABOVE As in aviation, red denotes 'remove before flight' to provide a visual cue to ground handling, test and support equipment. *(KSC)*

BELOW The Entry, Descent and Landing phase as planned for Spirit and Opportunity. *(JPL)*

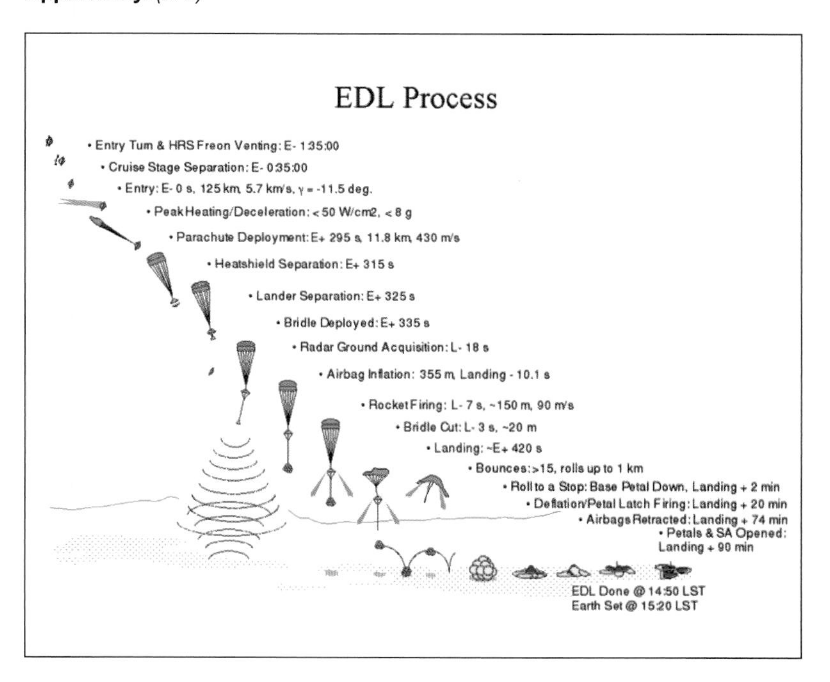

ABOVE A test rover demonstrates how Spirit and Opportunity moved from the lander to the surface of Mars. *(JPL)*

ABOVE RIGHT Spirit pauses to look back at its lander and the 'bat wing' safety panels. *(JPL)*

RIGHT With the rover gone, Spirit's lander is lifeless and inert, a monument to a successful mission. *(JPL)*

confirmation that the egress aids had been deployed and that there was no debris in the vicinity to foul roll-off. After these images had been taken the HGA would be aligned with the Earth for direct communications with the Deep Space Network. Before the rover started its programmed stand-up sequence, the petals would be moved to give attitude-sensing gyroscopes an idea about the stability of the surface, and if it was deemed acceptable the sequence to get the rover independent of the lander would begin.

As explained elsewhere, the rover was squished down on the lander base to fold within the limited volume created by the tetrahedron-shaped form of the enclosed petals. The sequence guided the rover's rocker-bogie system by first lifting the rover to allow the rockers to deploy, then lowering the rover on to the bogies. The rover would then be severed from mechanical connection to the lander and orientated to drive off in the appropriate direction, free of obstacles and with the petals to a hyper-extended position if necessary. It would take a lot of electrical power to go through all these mechanical procedures and they would take place only during daylight when the area was well lit.

The exact amount of time taken to go through what is in reality a complex series of go/no-go decision points, including the condition of the retracted airbags and the surface around the lander, depended upon a series of incremental steps, all of which would be documented for analysis by flight controllers after they took place – just in case some hazard awaited undetected by the Hazcams.

It would take up to four sols to get the rover off the lander and on to the surface, all the time inching its way along. Proceeding through what flight controllers call a go/no-go strategy, six decision points would define critical stages in the egress function. It would involve downlinking data from the rover, assessing whether the action had been carried out correctly and uplinking a series of commands to move to the next step. This cycle would take around 90 minutes. The decision tree required near faultless performance at each decision point but would ensure the rover was protected from undue electrical demands from the solar arrays and the batteries and from unexpected engineering problems.

After the rover had been moved to the surface under its own power, a calibration campaign would follow whereby each instrument would be checked and tested to ensure it was operating as expected. The lander/rover combination would have experienced a wide range of dynamic events and several months of operation in a difficult and hostile environment, and nobody would be able to get to it to check out equipment that for several months had been contained within the encapsulated structure, unattended.

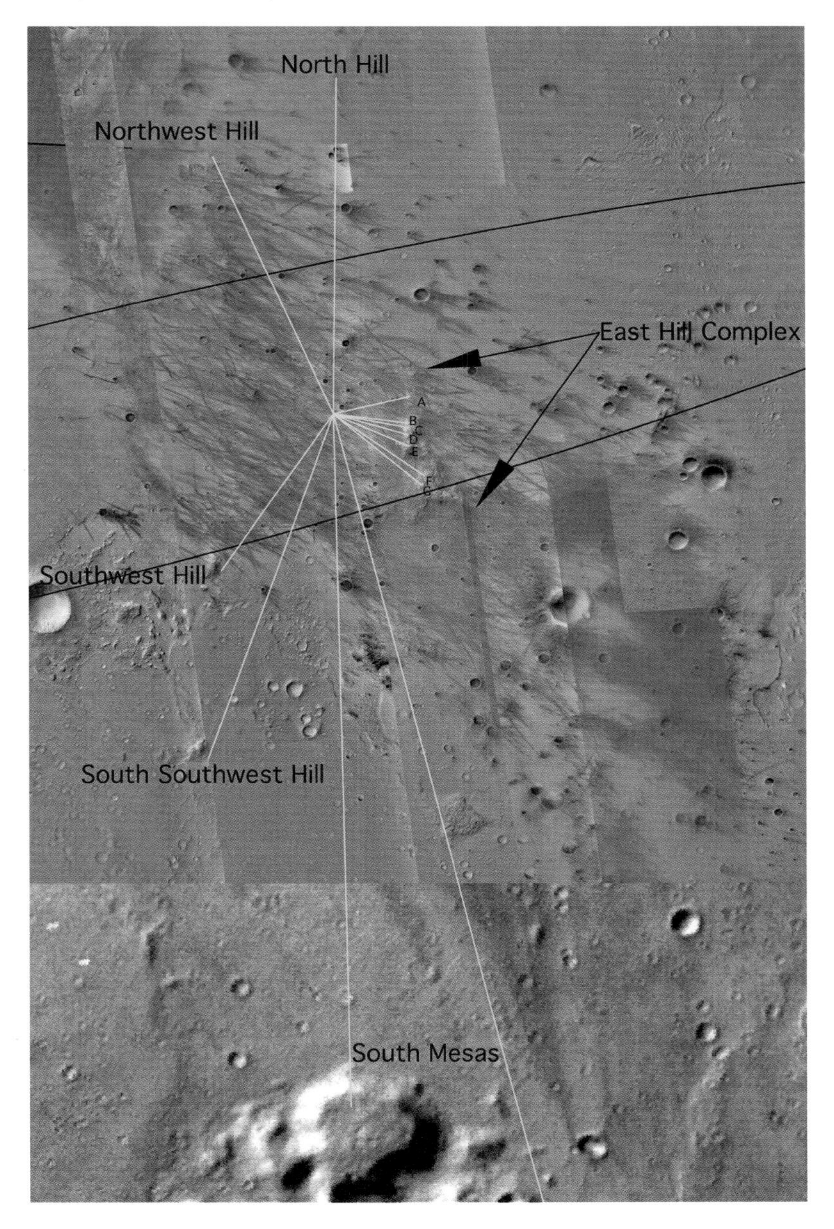

BELOW Scientists plot the location of Spirit and use panoramic camera views to plot precisely its position on the surface. *(JPL)*

One of the first tasks, conducted even before the rover left the lander, would be to establish exactly where the spacecraft was on the surface – not as easy as it sounds. From the last course correction down to the surface, the spacecraft would have operated autonomously all the way down through the atmosphere. A lot of things could happen to it during the EDL phase that could carry it several miles away from the planned landing spot.

A combination of post-landing imaging of the features around the spacecraft and a two-way Doppler VHF measurement between the rover and the Odyssey spacecraft passing in orbit overhead would help pin down the spot to within 300ft (100m). Detailed maps drawn from high-resolution imagery from Mars Global Observer and from Odyssey would allow scientists to match surface features seen from the Pancam views with those from orbit. This would have the advantage of allowing very precise sunset/sunrise times to be predicted for days and weeks ahead and for planning routes across the surface for the rovers. Unlike Sojourner, Spirit and Opportunity would roam across the surface beyond the visible range of their cameras, and correlation of route planning with images from orbiting spacecraft was essential to planning the best traverse routes.

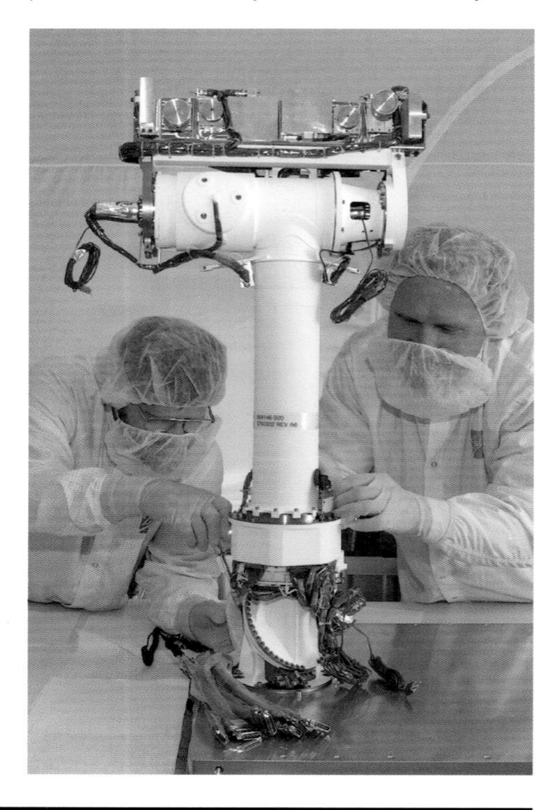

RIGHT The Pancam Mast Assembly carries the two Pancam cameras, the two Navcam cameras and the Mini-TES sensor platform that uses the PMA as a periscope. *(JPL)*

BELOW The Pancam optics are aligned to provide stereo pairs which will be useful for scientific measurements of surface features, as well as for engineering information when planning traverse routes. *(JPL)*

Science on the surface

The rovers each carried an identical suite of science instruments, including two panoramic cameras (Pancams), a Miniature-Thermal Emission Spectrometer (Mini-TES), a Microscopic Imager (MI), a Mossbauer Spectrometer, an Alpha Particle Spectrometer (APXS) and a Rock Abrasion Tool (RAT). The last four pieces of equipment were carried on the robotic arm attached to the front of the rover, which had an overall reach of 3ft (90cm).

The surface around the lander and rover would be surveyed by the Pancams providing visible and infrared images and by Mini-TES, which would generate data to characterise the nature and type of rocks in the vicinity by observing the thermal emissions spectroscopy. From the choices thus identified, specified rocks and surface features would be directly analysed by the Mossbauer Spectrometer and the APXS on the robotic arm. And for additional measurements of subsurface mineralogy, the RAT would be used to remove the outer layers of rocks and samples.

In addition, but critical to the navigation and autonomous operation of the rover, were four black-and-white Hazard Cameras (Hazcams), two at the front and two at the rear. These were the 'eyes' of the rover for all movement and navigation, providing engineers with adequate visual information to know where the rover was and to plan where it was going to go, as well as how it would get there. Operated in pairs, and mounted directly on the body of the rover, the Hazcams could map out a field of view 120° wide as far as 10ft (3m) in front or behind, with a width of more than 4m at the farthest distance.

Also known as the Instrument Deployment Device (IDD), the robotic arm has three flexible joints: shoulder, elbow and wrist. Five geared motors control movement, using these three articulated joints much like a human arm. The shoulder joint lies at the base of the attachment point and allows for 160° of movement in the lateral (side to side) plane and 70° of movement in the vertical (up and down) plane. The elbow joint is located midway along the arm and has its own dedicated motor moving the lower arm, wrist and 'hand' through 290°, to the folded or up and out position. The wrist joint incorporates two separate motors, one moving the suite of instruments it carries through a total range of 340° up or down and a second motor rotating the instruments through a spin circle of 350°.

The arm itself is operated in conjunction with the Hazcams, which were placed so as to take images of rock targets. The rover's computer positions crosshairs on to a target selected by ground controllers so that the arm could adjust the approach angle and line up with the target selected for analysis. Each instrument has contact sensors, dubbed 'curb feelers', to shut off the motors when the instrument makes physical contact.

Each MER rover had a single mast carrying the Pancam (panoramic camera) optics. This comprised two high-resolution stereo cameras forming the crossbar of a 'T' shape created by the camera mount and the mast. The instrument's narrow angle optics would provide images with three times the resolution of the camera carried by Pathfinder, and these would play a vital role in helping scientists decide the type of rocks to examine and to select the route to get to them. Working with the Mini-TES, the Pancams would image the surface

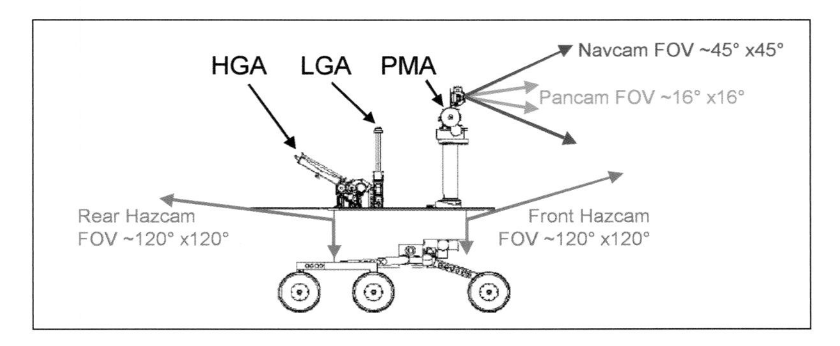

ABOVE The Hazcams are a vital part of giving the rover autonomy to proceed on the basis that if it meets an unexpected obstacle it will stop, while the Navcams are the primary optics for planning routes. *(JPL)*

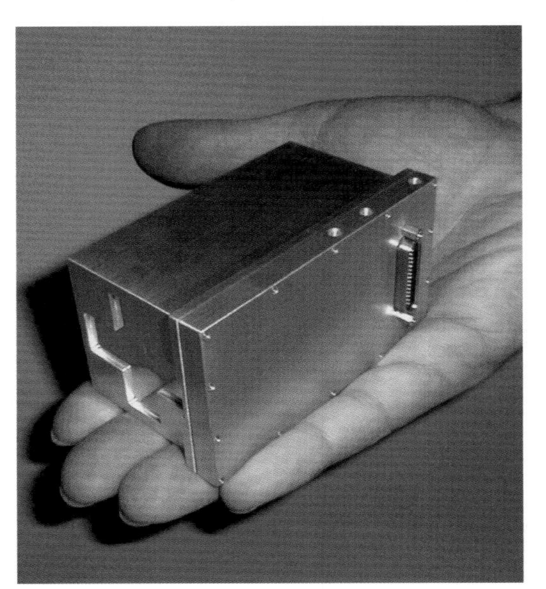

BELOW LEFT Supplied through a cooperative venture with the Max Planck Institute for Chemistry in Germany, the Mossbauer Spectrometer would search for iron-bearing minerals in rocks. *(JPL)*

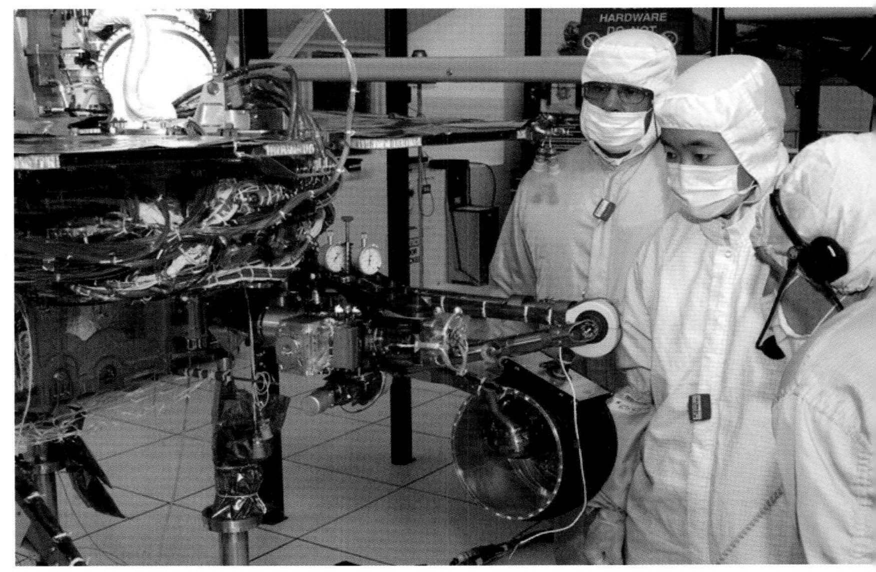

BELOW Contact science using the Mossbauer Spectrometer, the APXS and the Rock Abrasion Tool (RAT) would be conducted through the robotic arm, or Instrument Deployment Device, mounted to the front of the rover body. *(JPL)*

and its structures with unprecedented levels of information. Pancam measures light in 13 bands (430–1,100Nm) and in one clear band for far and near distance stereoscopic surface imaging. Blue and near infrared solar filters afford sky imaging capability.

Pancam has a 1,024 x 2,048 pixel Mitel CCD array detector operating in the frame transfer mode, one 1,024 x 1,024 pixel region constituting the active imaging area and another adjacent region with the same pixel quantity serving as a frame transfer buffer. The transfer buffer has an opaque cover that stops more than 99% of the light at all wavelengths from 400 to 1,100Nm. The arrays are capable of exposure times from 0msec to provide a 'readout smear' signal acquired during the ~5msec necessary to transfer the image to the frame transfer buffer. Maximum exposure time is 30 seconds. Analogue to digital converters provide a digital output with 12-bit encoding and a signal-to-noise ratio greater than 200 at all levels above 20% of full scale.

Each array is combined with the optics and a small filter wheel, and the optics for both cameras consist of identical three-element symmetrical lenses with an effective focal length of 38mm and a focal ratio of f/20, yielding a square field of view of 16.8° x 16.8° per eye. A sapphire window located at the front of the optics barrel protects the optics and the filters from direct exposure to the Martian environment. The design of the optics maintains optimum focus from infinity to within approximately 4.9ft (1.5m) of the cameras. At closer range, Pancam is progressively out of focus so that, for instance, at a range of 31.5in (80cm) the blur is about 10 pixels. The filter wheel has eight positions for each optical set, the left containing the one clear position, the remainder filled with narrowband interference filters, circular and 0.4in (10mm) in diameter.

The two Pancam cameras are separated by 12in (30cm) horizontally and have a 1° toe-in which together provide an adequate convergence distance for stereoscopic imagery and for ranging operations, the effective distance for these being 16.4–32.8ft (5–10m) to approximately 328ft (100m). The pointing control is less than 2° in azimuth and 1° in elevation. Pancam operation would usually take place during daylight to receive high-quality measurements of sunlight reflected off rock samples and soil, but the technical capabilities of the cameras would permit twilight images of astronomical objects and night-time subjects. The design of the cameras gave them an optimum operating temperature of –55° up to 0°. The boresight of the cameras was about 4.3ft (1.3m) above the surface when the PMA was fully erect and the camera assembly could move in +/–90° elevation using a geared brush motor on the horizontal camera bar and 360° in azimuth.

The Pancam Mast Assembly (PMA) also included several key pieces of equipment for the Mini-TES, an instrument that had been a key element in the design of the Athena rover when its mission, and its launch date, had been undecided. The Mini-TES is a Michelson interferometer mounted inside the rover, viewing the surface by using the PMA as a periscope. It uses a scan mirror to reflect radiation down the PMA tube into the telescope and the interferometer to produce scan cubes over a 360° range in azimuth and from –50° to +30° in elevation. To protect the optics from dust, the mirror could be moved to a stowed position where a cover blocked the aperture. Mini-TES measures the emission spectra of rocks and soils in 5–29μm. The Mini-TES structure weighs 5lb (2.1kg) and is located within the main body of the rover just below the Pancam Mast Assembly.

To obtain information about the mineral composition of the terrain around it, the Mini-TES was located on the top of the Pancam Mast Assembly (PMA) so as to 'see' the same view as the Pancam. When acquiring data, the PMA's elevation mirror and azimuth actuator are sequenced to generate a raster image of the scene. The elevation and azimuth servos could move to a commanded position with an accuracy of 1mrad. The telescope at the base of the PMA is a reflecting Cassegrain type with a 2.5in (6.35cm) diameter mirror and a focal ratio of f/12. The linear motor drives are the same as those used on the Mars Observer and Mars Global Surveyor spacecraft. The instrument would usually be used during the midday period from 10:00am to 3:00pm local Mars time but night-time observations would be possible to measure surface and atmospheric temperatures on the full diurnal cycle.

Until deployed on the Martian surface, the Pancam Mast Assembly would be stowed across the top of the rover equipment deck (RED), attached at the forward end to a firm mounting plate which permitted it to be rotated 90° to a vertical position. Two locks at either end of the Pancam bar would ensure it remained firmly attached without threat of movement during the dynamic events associated with landing. In the stowed configuration it lay underneath the cylindrical parachute container, which was itself attached to the underside of the backshell. When the backshell was jettisoned the parachute it contained would go with it, freeing the space above so that when the lander petals unfolded the PMA could be erected.

This would be achieved by an actuator at the base, triggered by the firing of a pyrotechnic restraining bolt. Pyrotechnic devices were developed during the Apollo programme as a safe and simple way of disconnecting bolts and strong tie-down devices in the vicinity of people. As such, and because they collect the explosive debris rather than releasing it as microscopic debris, they are ideal for operation in close proximity to sensitive instruments or electronics. Once released the PMA arm would sweep out in a circular motion and be permanently locked in the erect position and could not be folded back down.

The entire PMA head can be rotated 360° in azimuth by a geared brush motor assembly. A separate, dedicated geared brush motor controls elevation of the Mini-TES elevation mirror assembly. The two navigation cameras (Navcams) are also located on the same camera bar and point in the same direction as the Pancam cameras. The two Navcams could produce black-and-white images in visible light, gathering panoramic, three-dimensional pictures with a 45° field of view to support navigation planning for routes and destinations across the surface.

In addition to the two Pancam cameras, the two Navcam cameras and the four Hazcam cameras, the rover also carried a monochromatic Microscopic Imager (MI), attached to the robotic arm. Its primary function was to provide geologists with close-up images of small rocks, pebbles and dust. Essentially a combination of microscope and a CCD camera, it would work in conjunction with other instruments to study exposed sections of rock scoured out by the Rock Abrasion Tool. It was to prove invaluable in determining grain size and the shape of micro-grains of sand to establish whether they had been part of a sedimentary deposition and whether water may have been involved.

LEFT Carried on the front of the robotic arm, the APXS sensor head would be used to obtain spectral readings from rocks, with the data carried through wires to the instrument inside the body of the rover. *(JPL)*

The MI has a simple fixed focus optical design at f/15 which provides +/– 3mm depth of field comprising 1,024 x 1,024 pixels; with a focal length of 20mm and a working distance of 63mm, it has a field of view of 32mm x 31mm. It has a single broad-band filter for its black-and-white images and uses only natural light, but stereo pictures can be obtained at a variety of distances through multiple images, and also by several different images of the same target taken at different distances. Mounted to the robot arm (IDD), the MI controls accurate positioning relative to its target through a special

BELOW The RAT cleaned rock surfaces and scoured the surficial coatings to get inside the rock for APXS and Mossbauer measurements. *(JPL)*

RIGHT Spirit applies the RAT to get into a rock sample in Gusev Crater. *(JPL)*

FAR RIGHT The RAT tool guard was fabricated from a piece of metal recovered from the debris at the site of the World Trade Center in New York, destroyed by terrorist actions on 11 September 2001. *(JPL)*

contact sensor and once in position requires only 15 seconds for motion-damping.

To grind down the outer surface of selected rocks and surface materials, the robotic arm was equipped with a powerful grinder, the Rock Abrasion Tool (RAT), which could create a hole 2in (4.5cm) in diameter and 0.2in (5mm) deep. With a weight of 1.6lb (685g), the RAT measures 2.7in (7cm) in diameter and has a length of 4in (10cm), drawing only 11W of power for its operation. It is equipped with three electric motors to drive two sets of grinding teeth, two of which rotate at high speed with the capacity to rotate around each other at slower speed, so sweeping the entire area. The RAT was capable of grinding through hard volcanic rock in about two hours. In this way, geologists would be able to get beneath the outer layers of material affected by weathering or surface conditioning, a layer that could be very different from the nature of the rock itself. Only by breaking through that layer could scientists fully understand the history of the rock itself.

BELOW RAT holes made by the abrasion tool provide access to rocks covered by several millennia of deposition and surface chemistry modified by weather effects and dust. *(JPL)*

The Mossbauer Spectrometer is a two-channel elemental chemical analyser with a 0.6in (1.5cm) diameter contact sensor and dust door. It has a sampling depth of 200–300μm and was primarily in search of the composition and abundance of iron-bearing minerals in the rocks found at the surface. The sensor head is located on the robotic arm, with the electronics situated down in the rover's Warm Electronics Box.

Measurements would be made by placing the sensor head directly in contact with the rock or soil sample with a contact plate at the front of the head to a distance of about 0.35–0.39in (9–10mm). A heavy metal collimator at the front of the source irradiates a spot about 0.6in (15mm) in diameter on the surface of the sample. The average depth of sampling by this method is 200–300μm. Because the measurements are highly sensitive to temperature, a wide range of measurements would be made over a full Martian 'day', obtaining data from maximum to minimum ambient temperatures. The instrument and its electronics package were supplied by the German Space Agency and by a small team at the University of Mainz.

The German Space Agency and the Max Planck Institute for Chemistry also provided the APXS instrument. It was included in the suite of scientific instruments to use X-rays for determining the elemental composition of rocks. The sensor head was mounted on the robot arm, but the electronics were down in the comfort of the Warm Electronics Box where they could be protected from the extremes of the Martian environment. The sensor head contained six curium-244 sources arranged radially around the sensor head with a total strength of approximately 30mCi. The head comprised a cylinder 2.6in (6.6cm) in length and 1.6in (4cm) in diameter.

Collimators in front of these sources determined the instruments' field of view, which was about 1.5in (38mm) at a working distance of 9.5ft (2.9m), protected from dust by small doors. The inner surfaces of the doors provided a calibration reference surface. It had a sampling depth of 1–100μm and performed an identical function to the APXS carried by Sojourner, taking at least ten hours to get a good sampling. The APXS on the two MER rovers was a direct descendant of the instrument planned for the early Athena rover, which in fact became Spirit and Opportunity.

Each rover had three sets of magnets provided by the Netherlands to collect dust for analysis. Scientists believed that the magnetic minerals carried in fine particles could be preserved from the period when Mars had a watery climate, and their presence – plus the pattern of accumulation on the magnets – could point to clues about the planet's past environment. One set was attached to the front of the rover body for collecting airborne particles and a second set was attached to the top of the rover equipment deck. A third set was attached to the Rock Abrasion Tool to collect fine particles driven off rock surfaces by the grinding action of the tool. Magnets on the front of the rover and on the RAT could be accessed by the Mossbauer Spectrometer while that on the top deck was sufficiently strong to deflect wind-carried particles.

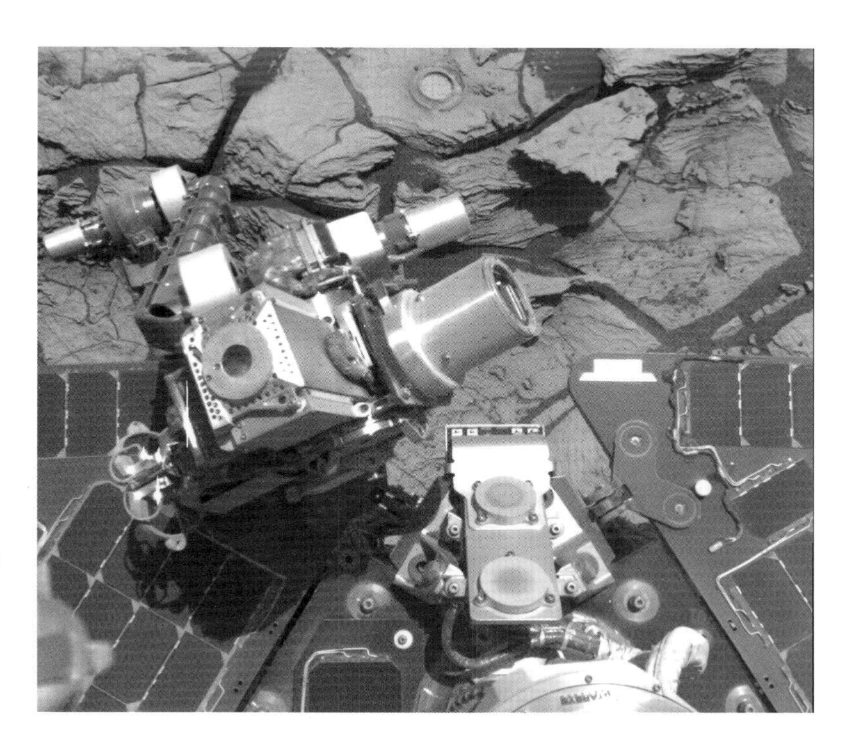

ABOVE Opportunity repositions its robotic arm for contact science in the Meridiani Planum. *(JPL)*

BELOW Even the action of the rover's wheels can be a scientific test of surface conditions, their action on the surface scouring out white surface deposits similar to ice. *(JPL)*

ABOVE **Virtual techniques for navigating rovers across the surface of Mars using computer simulation, pioneered by Sojourner, were considerably upgraded by flight controllers now supporting two advanced rovers on the surface of Mars.** *(JPL)*

At work on Mars

The science package of instruments and equipment aboard the two rovers was the responsibility of Dr Stephen Squyres, professor of astronomy at Cornell University, Ithaca, New York. His involvement dated from the days when he developed the Athena rover prior to its metamorphosis into Mars Exploration Rover. Getting the science team working as a coherent unit was virtually impossible, each investigator fighting, sometimes fiercely, for his or her own particular experiment. It was at best a compromise. That process of prioritising and integration begins at the start of a programme, long before final approval, and never ends until the mission is over. Scientists fight long and hard to get their own particular set of equipment on board; and after approval, when the mission reaches its destination, the fight switches to a bid for time, both on the surface and in the data flow back to Earth.

As each 'day' dawned on Mars, the spacecraft were set upon a course of action that had been planned over the preceding day, commands sent up by tracking stations of the Deep Space Network. The way the days are budgeted is a deft manipulation of a fixed timeline. The spacecraft will perform experiments coded into the computer commands transmitted in the hour preceding the event. For that slice of the timeline the different participants must argue their case at planning sessions where the decisions are made as to where to emphasise the science and what instruments should be integrated with the overall objective.

When the data is acquired it is often stored for transmission only when the telemetry and data flow allows sufficient capacity for it to be downloaded. In the case of Spirit and Opportunity, that can be through scientific

RIGHT **The seven Columbia Hills at Spirit's Gusev site were named after the seven crew members of the Shuttle *Columbia*, lost on 1 February 2003.** *(JPL)*

RIGHT The first major task for Spirit after touchdown was to assess the local area, but flight controllers quickly decided that Columbia Hills were sufficiently interesting to plan a concerted effort to get there before the rover failed. It would outlast by many years the limited expectations of its survival! *(JPL)*

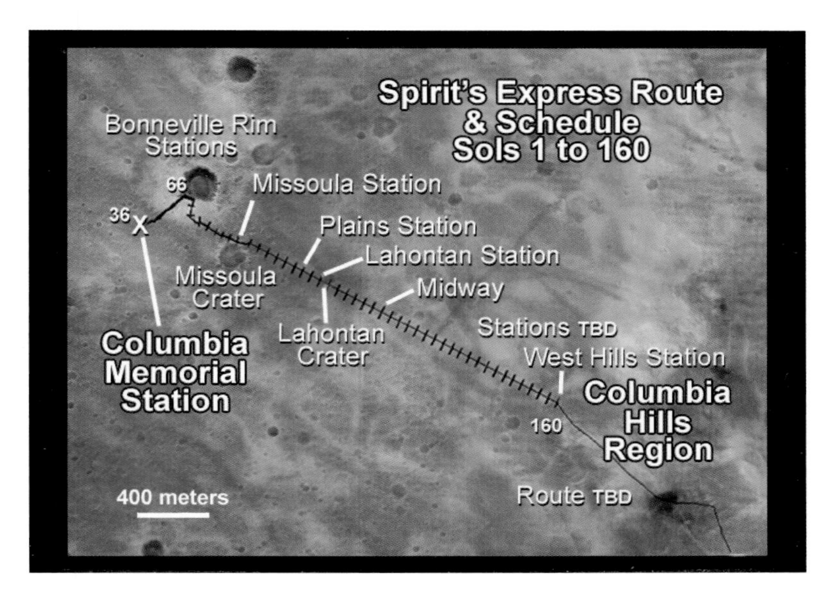

satellites orbiting Mars or direct to Earth. The equipment on board the rovers has been specifically designed to accommodate that, but the fight back on Earth for space in both time and data capacity is a fiercely contested line of priorities!

In space jargon, the word 'experiment' does not imply an experiment performed in the classic manner, where a test is made and a control or analogue applied to set it against a comparator. The word actually means an instrument or suite of sensors gathering data or information from which scientific measurements can be obtained. There has always been a slightly abrasive disregard for the word 'experiment' in terms of what a spacecraft does, and some scientists avoid the use of this word when referring to their work!

Managing people, whether as individuals or in teams, and integrating scientists, engineers and technicians is a difficult task at best, and on flights where operations roll along on a continuous basis the demands on time and energy can be crushing, both on body and mind – so much so that NASA pays some attention to evaluating the impact on people as they work 24 hours a day, seven days a week, very often without rest.

Some people have to rent apartments far from home to work from offices and control centres at JPL, only seeing their families once every few days or even weeks. When Mars missions are under way, the day-by-day planning and execution of schedules, timelines and tasks are tough and demanding, with the ceaseless and uncompromising need for accuracy and fault-free calculations, computer instructions and transmissions to and from the spacecraft – each 'conversation' with the latter taking up to 40 minutes between asking a question and getting the answer.

Unlike Pathfinder, each MER rover was to be capable of traversing up to 330ft (100m) each day, but that represented a maximum pace of drive with no stops for science or operation of any of the onboard instruments. Human operators could not plan courses longer than 66–98ft (20–30m) using imagery based on rover cameras, so an autonomous navigation system was essential. Sojourner's laser-projected stereo system produced 25 measurements for controllers to plan each step of the journey. MER would use sunlight to take up to 5,000 free-range measurements for each step.

BELOW Less than two months after landing, Spirit had made it to within 2,800ft of Husband Hill, the first 'target' in the Columbia Hills region, intermediate sites being awarded local names for planning identification. *(JPL)*

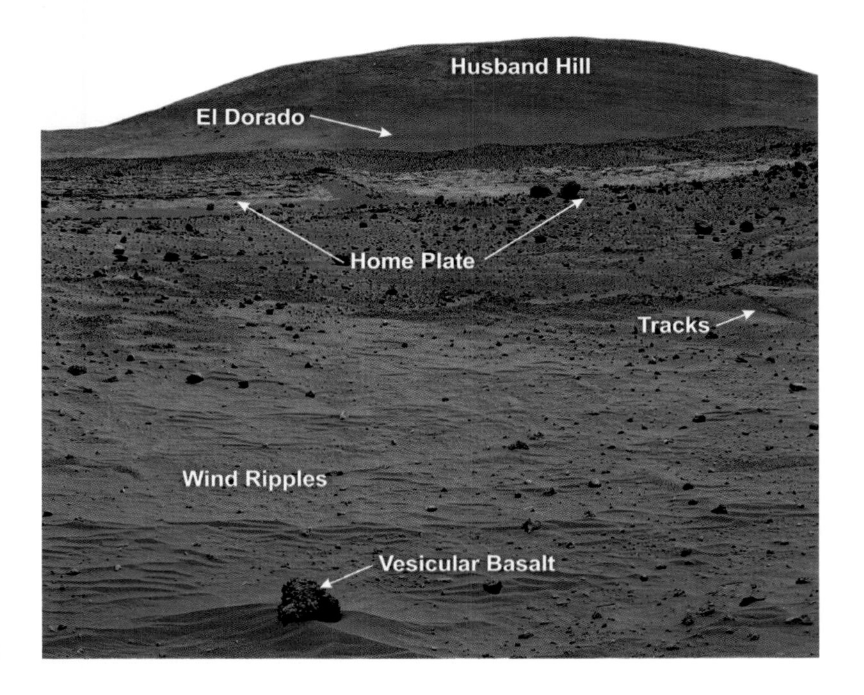

Using an image taken by Spirit and a 'virtual' rover superimposed in scale, Spirit pauses at Husband Hill to give a surreal effect of the intense isolation of the Martian vista. *(JPL)*

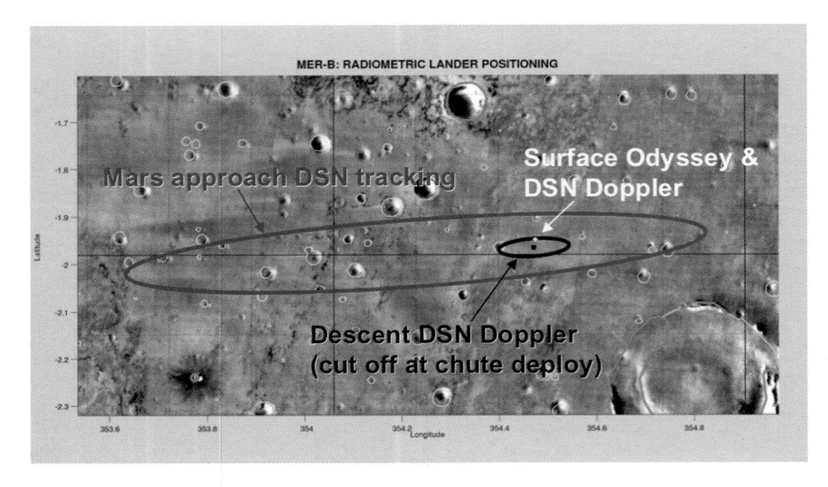

ABOVE The landing site for Opportunity is plotted, the large black ellipse being the predicted Doppler tracking solution from the Deep Space Network during the approach phase, the smaller black ellipse defining the DSN solution during descent, and the white cross being the actual position from images taken by the orbiting Odyssey combined with Doppler tracking of the signal from the rover. *(JPL)*

The ability to traverse the surface would be derived directly from these measurements via a map updated and maintained on board each vehicle. These allow each rover to choose its next move with some degree of intelligence, avoiding obstacles as it moves forward. In some respects this is similar to the terrain-avoidance capability in cruise missiles, reading a map of where they should be and correcting any diversions or deviations to get back on course.

One of the principal engineers involved with the design of this system, Dr Mark Maimone was employed at JPL as a machine vision researcher and had delivered 3-D visual systems to several test vehicles and operational systems, including one employed for visually inspecting the debris of the Chernobyl nuclear reactor when it suffered catastrophic meltdown in April 1986. Technology such as this is part of an integrated software-sharing function now widely used in 3D television and in cinematic productions but can also turn 2D images into 3D on demand.

Planning surface activities over the 91 sols of the primary mission depended upon several fluctuating 'windows' of opportunity. With Mars spinning on its axis in 24hr 40min, and with the rovers powered by solar cells, activity could take place over only half its surface time – during daylight when sunlight illuminated the solar panels. Communication with Earth could also only take place when the landing site was in view of the Earth or when an orbiting spacecraft could act as a brief relay as it passed overhead.

Although the MER missions were expected to see each rover trundling around on the surface for 91 sols, scientists hoped for much longer, although the threat of dust covering the solar-cell panels and cutting electrical power production below the level needed to keep each rover alive was problematical. But there was one further challenge to rover operations. The primary mission would end in late April after the rovers had completed their three months of exploration after landing, but toward the end of August communications with Mars would cease for a period while the planet passed behind the Sun as viewed from Earth. These periods of so-called conjunction, occurring every 780 Earth days, would always cause a brief hiatus in communications but were just one more consideration for a mission. Before then, around late June, the onset of winter would bring shorter days, less energy from the Sun and much lower temperatures for about six months before the onset of spring.

A typical rover day begins at sunrise or when the onboard alarm clock goes off. The alarm clock is a better marker to start the day than the detection of sunrise because the time when solar arrays start producing power is dependent on opacity of the atmosphere, shadows, dust and a range of imprecise values. During the X-band communications pass, commands for the day are uplinked to the rover, which

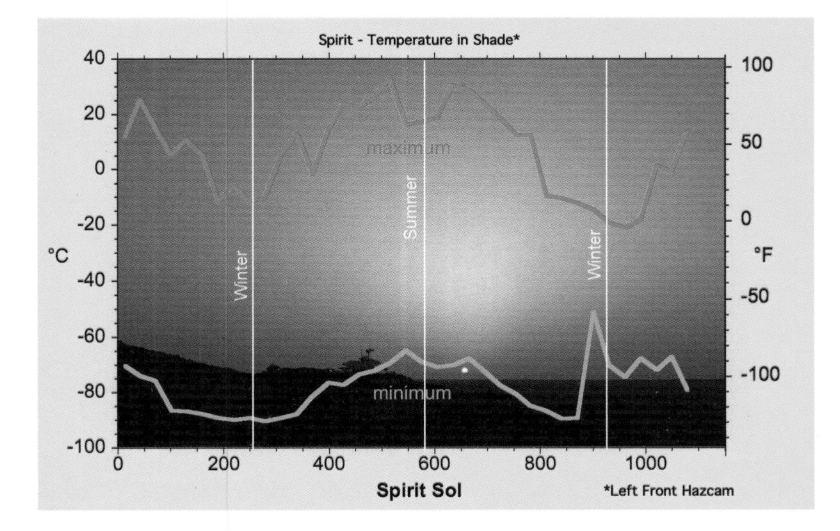

LEFT Spirit sends continuous temperature data, this chart showing variations according to night and day and the seasons of a Martian year. *(JPL)*

then activates the master sequence by first deactivating the previous command sets – so as not to confuse the computer as to the assigned tasks and attempting to control the rover through two sets of instructions!

The uplinked instructions programme the rover to conduct a precise set of activities, be they driving across the surface, stopping to investigate a rock or surface feature, or simply remaining in place to continue investigations with selected instruments. The rover will usually take a midday siesta and wake up on its alarm clock going off, resuming the master sequence already in the computer. During the afternoon X-band pass the rover downlinks data, but occasionally, if there is a link, flight controllers will upload additional computer segments.

It is during the afternoon session that flight controllers get the data they need to plan the next sol, and this is when the scientists and engineers go into a huddle to determine priorities and activity for the following sol. During the afternoon and early in the morning, UHF links via the orbiting spacecraft will provide greater flexibility for mission operations. Planning activity is different every day and a constant dialogue occurs between the engineers monitoring rover systems and advising on what is possible for the next sol and scientists deciding how best to use the resources of the rover for conducting the most beneficial science.

The greater the demands of the scientists, the more problems there were for the engineers. The Martian environment changes dramatically over time as the planet migrates through its 'year', the period of time it takes to go around the Sun, which is equivalent to 687 Earth days. Mars is tilted on its axis a little over 25°, only a little greater than Earth's 23.5°, and because of this its seasons are similar to ours. But the orbit of Mars is much less circular than Earth's orbit and this brings great fluctuations in seasonal variations.

Because of this greater eccentricity, autumn and winter in the northern hemisphere are

BELOW A view from a Navcam on Spirit as it approaches West Spur on 29 July 2004 to view the outcrop indicating water flow. Top right is an expanse of basaltic lava that long ago spread out across the area toward Husband Hill. *(JPL)*

Opportunity

“Cape Verde”

“Duck Bay”

“Cabo Frio”

much shorter than the spring and the summer seasons, while in the southern hemisphere the opposite is the case. In respective hemispheres, summer brings atmospheric stability because the poles are warmed by continuously pointing (respectively) at the Sun. As on Earth, winter storms are caused by fluctuations in temperature, but because the poles, in their respective winters, are in a state of continual darkness the temperature of the primarily carbon dioxide atmosphere plunges below the freezing point of CO2 (–123°C).

When this happens the CO2 freezes out, creating a great expanse of ice – not of water as it would on Earth, but of carbon dioxide. One-third of the atmosphere condenses down on to the surface in respective winters. But as one pole is condensing and cooling down, the other is evaporating under the warm Sun. Large circular motions of air stirred up by these fluctuations cause enormous circulation patterns to develop, bringing surface changes observable from Earth.

It was these seasonal changes, and the great dust storms that can develop quickly and envelop almost the entire planet, that brought major threats to surface rovers, creating problems for flight controllers fighting to keep the vehicles alive. Dust settling on solar-cell panels quickly degrades the production of electrical energy, the lifeblood of any spacecraft dependent on this source. And winter, with low Sun angles and short days, limits the amount of work a rover can do as power levels fall.

The seasonal changes on Mars have been known about for centuries, and the 'wave of darkening' now known to be caused by high winds and dust storms are not the signs of the ebb and flow of vegetation but merely the effects of an annual cycle that has been going on for millions of years. When Mariner 4 took the first close-up pictures during its fleeting fly-by in 1965, the craters and lunar-like appearance of the surface was a great disappointment to many.

Over the decades since, Mars has become increasingly attractive as a place to study how other worlds evolve and establish their own climatic and geophysical characteristics. There was only so much that could be done from orbit, and the surface observations of the two Viking landers were insufficient for a full understanding of the planet. Vehicles equipped with scientific instruments and capable of roaming the surface of Mars were essential to a more complete understanding of the planet. But nobody could have foreseen the phenomenal performance of Spirit and Opportunity after they landed at their respective sites in early 2004.

Mission outstanding

Although the two rovers and the two landers had been designed and put together at JPL in California, when they arrived in Florida for launch they would go through a rigorous test and final assembly in the Payload Hazardous Servicing Facility (PHSF) operated by the Air Force at Cape Canaveral Air Force Station. The process through which each spacecraft would go was a management and engineering flow known as ATLO – Assembly, Test and Launch Operations. This phase begins when NASA is sure that everything is on track as far as predictably possible. For MER it started in January 2002 when the pre-ATLO reviews were held. This was when serious doubts arose over cost and the readiness of the scientists and the engineers to make it all the way to the pad for launch in 2003. But the real ATLO process would only start with delivery of the two spacecraft to Cape Canaveral.

OPPOSITE Opportunity at Victoria Crater, a depression 2,500ft in diameter and 250ft deep, identified from an image taken by Mars Reconnaissance Orbiter in 2006. *(JPL)*

BELOW Opportunity takes a Pancam view of a meteorite sitting on the surface, shown here in false colour to highlight the different minerals. *(JPL)*

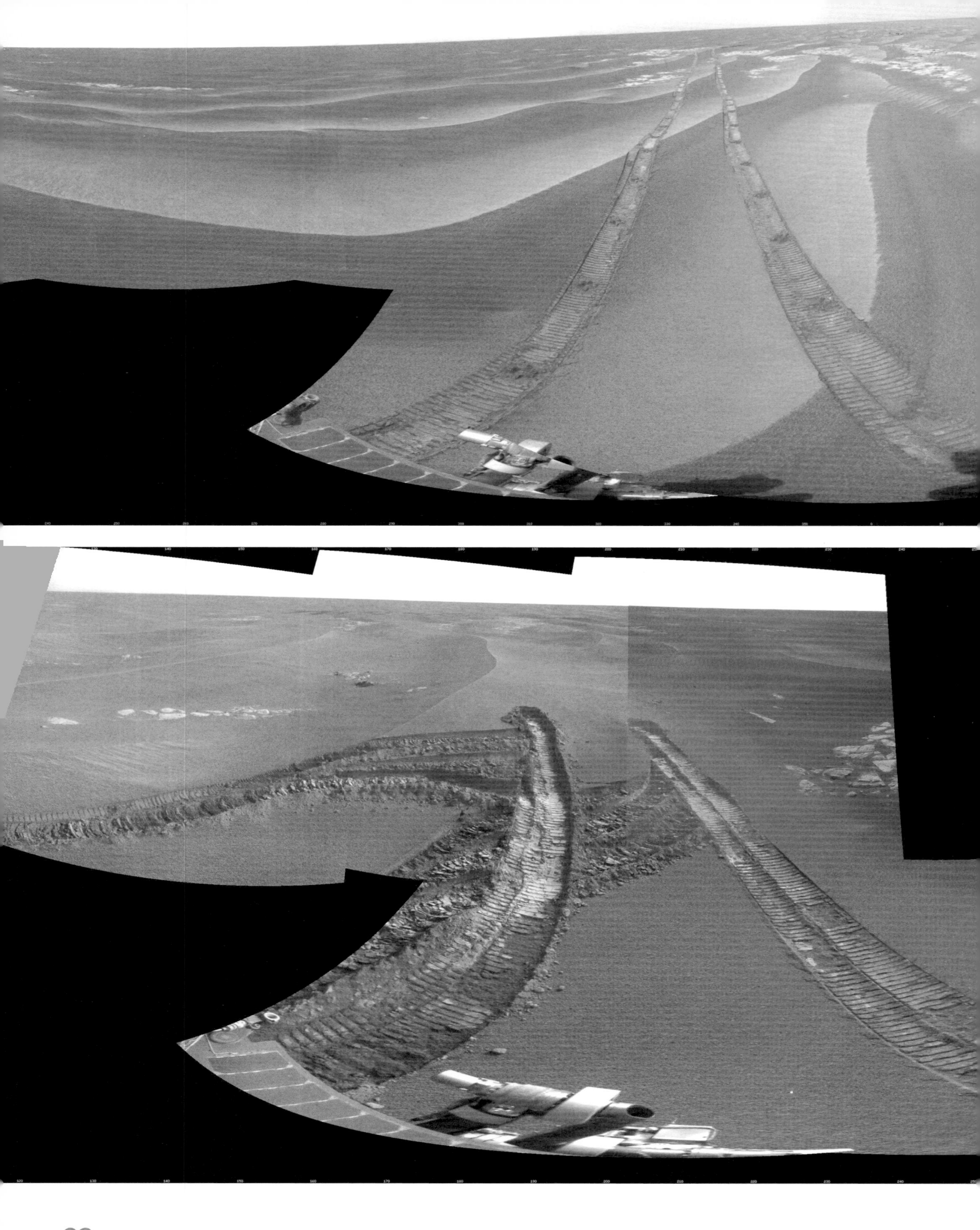

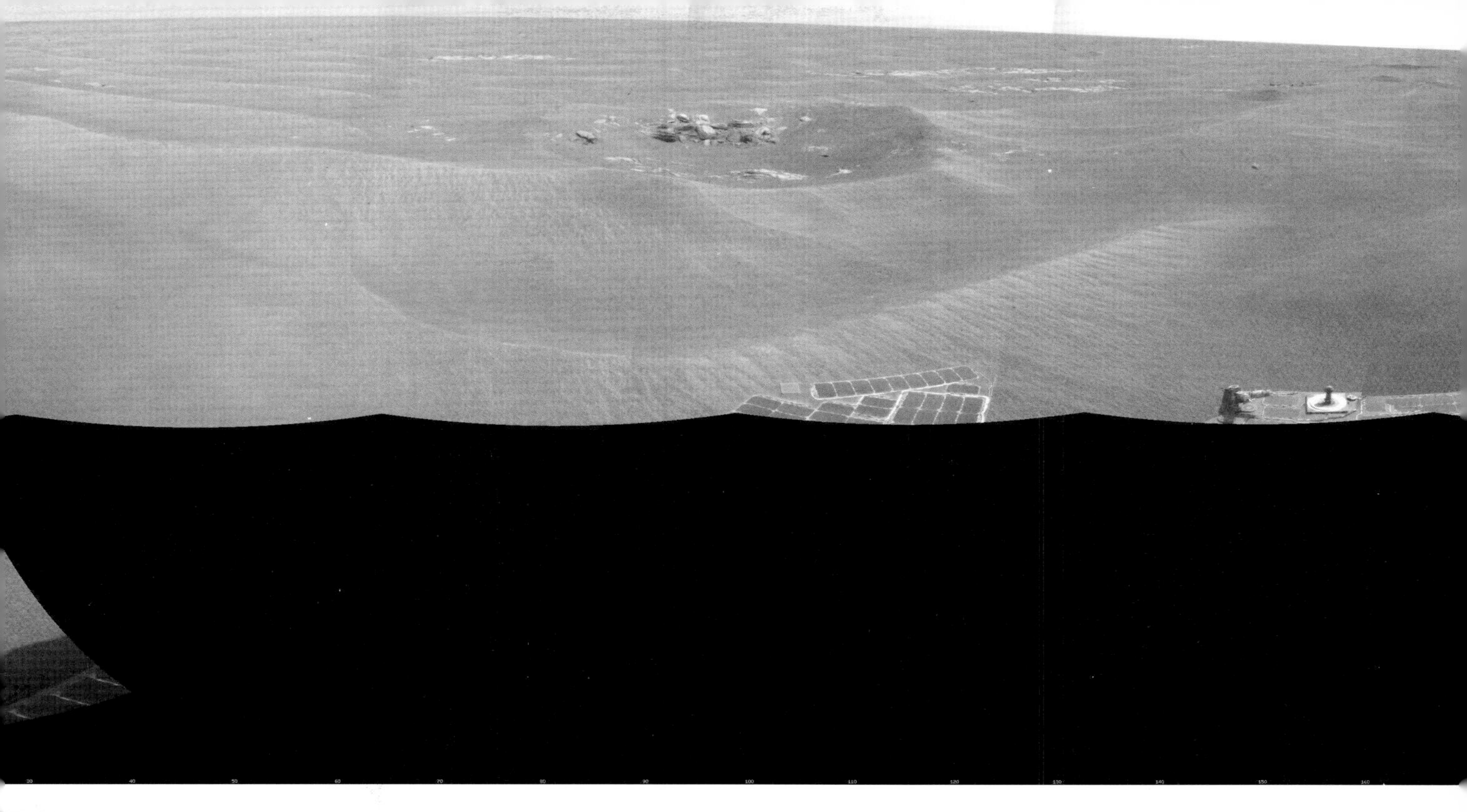

As work to get the first spacecraft (MER-2) trucked across the United States to Florida reached its final stages, a shocking event was unfolding in the skies above the southern states. On 1 February 2003 the Shuttle *Columbia* broke up attempting to return to Earth after a lengthy science mission, killing all seven crew members. This was the second Shuttle accident following the loss of *Challenger* and another seven lives on 28 January 1986. *Columbia* is immortalised on Mars by the naming of a particular feature as the Columbia Hills.

There was another association with a national tragedy when word of the destruction of the Twin Towers in New York in the 11 September 2001 terrorist attack reached the MER team. They arranged for a tiny piece of aluminium recovered from the collapsed site of the World Trade Center to be incorporated into Spirit. It was manufactured in as a cable guard on the Rock Abrasion Tool at the end of the IDD.

MER-2 left Cape Canaveral by road on 22 February 2003, and other items of equipment also began to make their way to Florida, including the APXS and the Mossbauer Spectrometer that had been contributed by Germany. But when pre-assembly tests began of all the many separate elements, word from Germany brought concerns that the APXS

ABOVE Rover tracks across the dunes from Opportunity after it had been driving backwards to relieve the strain on the front right wheel which drew more current than usual, indicating low lubrication levels, in April 2009. *(JPL)*

LEFT Opportunity backs out of trouble, reversing away from a sand-trap into which it was heading and from which it would have been unable to get free – a fate which befell Spirit. *(JPL)*

instrument could be flawed. That necessitated some replacement with elements already held on standby, but they had to be hand-carried to JPL before rushing to Florida for installation aboard the rover.

As tests continued and integration went forward toward the early 8 June launch date, MER-1 (Opportunity) fell behind MER-2 (Spirit), so the latter was selected to go first as Mission A. Henceforth the public would know them only by their names and not by the confusing and misrepresented assignment of the second rover to the first mission! Then, in mid-May, a potentially catastrophic electrical short occurred in the Telecom Services Board located deep within the Warm Electronic Box. With only three weeks left before launch, and Spirit's solar panels already folded and stowed, this was no time to be digging back down deep inside the rover.

When fuses blow during tests it can be due to any one of several terrifying reasons, polarised around a fault within the spacecraft or a fault in the GSE (Ground Support Equipment). The GSE test equipment brings a spaghetti-tree of cables, lines and electrical wiring plugged in to the spacecraft as though it were a patient in an operating theatre. Sometimes the GSE equipment can overload delicate spacecraft circuits for the most arcane of reasons – two cables can be too close and create an electromagnetic arc that significantly boosts current beyond the level the system is designed to cope with. Such mistakes should never happen but they do. It was the result of using incompatible GSE equipment months before launch that nearly doomed the Apollo 13 crew

BELOW Spirit appears covered with wind-blown dust, seriously eroding electrical production from solar cells unable to get essential sunlight. *(JPL)*

in deep space following an explosion on the way to the Moon in April 1970.

After careful analysis, looking at the test schedules, the electrical loads applied to each test and to each pyrotechnic device, the metals out of which the relays had been manufactured and the duration of the current in milliseconds, the reason became apparent. The metal used on the solar array pyrotechnic cable-cutter included zirconium, which held a charge after firing of 32 milliseconds, whereas the fuse would function at less than 1 millisecond. To release the MER rovers for launch, NASA had to be convinced that all the other pyrotechnic devices would work as planned and not be inhibited by trigger-happy fuses, which in reality work as circuit-breakers.

By 21 May the problem had gone all the way up the chain of command to NASA headquarters and there were serious doubts that the mission could fly that year. Any delay would flip the launches into 2005, when the next window of opportunity would come around, and that was an impossibly bad window for the delicate balance between the weight of the two spacecraft and the power of the two launch vehicles selected for the mission. Most of the HQ management team, including Ed Weiler, head of science projects, were convinced the flights would have to be delayed but they agreed to keep going to the Mission Readiness Reviews, the first of which was planned for 30 May – eight days before the launch window opened at Cape Canaveral.

Launch windows occur at intervals when the planets are in the best position for a flight

BELOW More than a full Martian year after it arrived at Gusev Crater, Spirit uses multiple images to create a mosaic showing its solar cells cleaned again after the wind had blown stronger. *(JPL)*

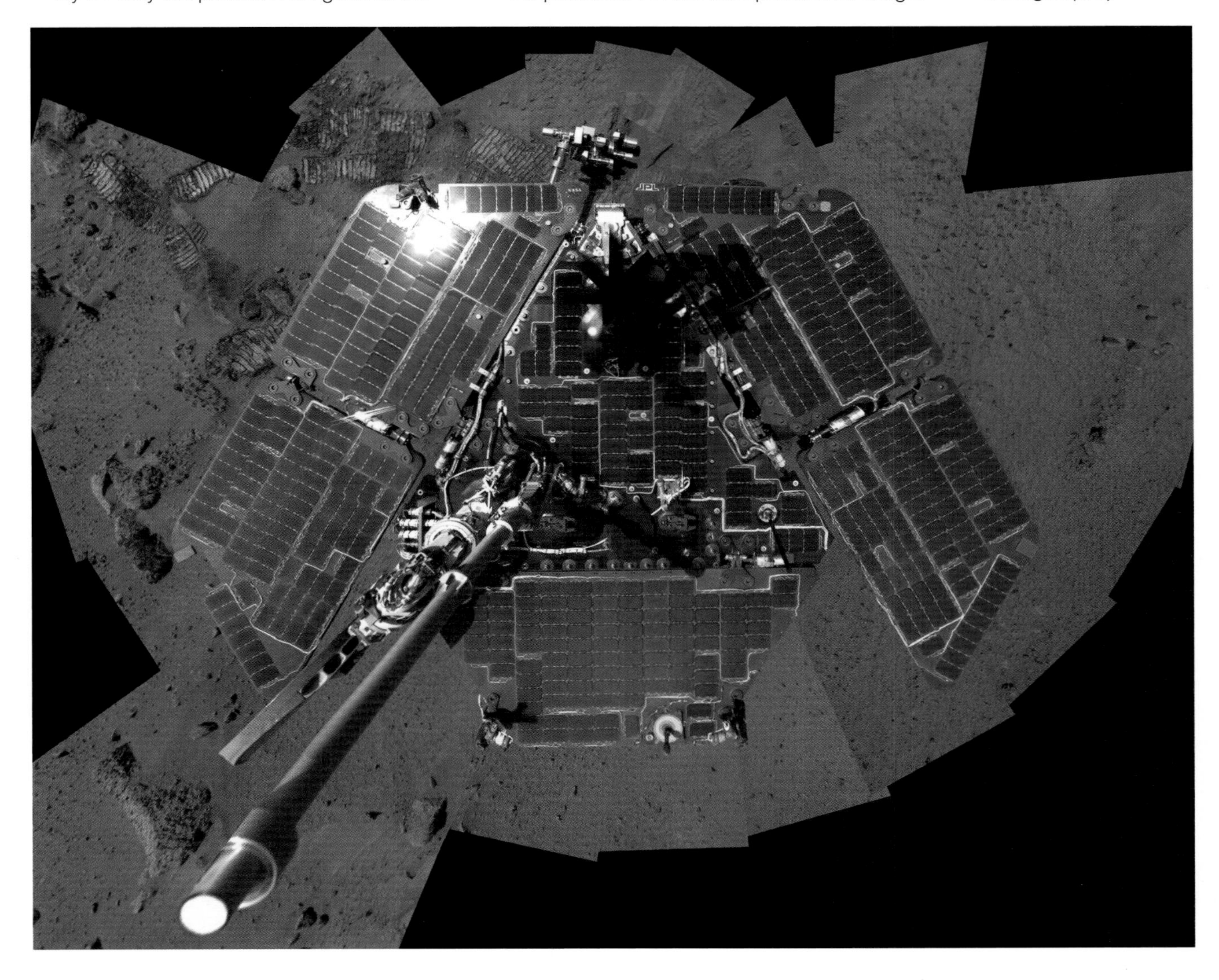

between Earth and Mars, and as we have seen that is roughly once every 26 months. But the amount of energy needed to send a spacecraft to Mars in these windows varies according to the relative position of the two planets in their elliptical paths about the Sun. In late summer 2003, for instance, Earth and Mars were on the same side of the Sun and on 28 August they were the closest they had been to each other for 59,619 years – a 'mere' 34,646,418 miles. This was a particularly good launch window and allowed energy savings in propelling the two spacecraft to the Red Planet. A delay to 2005 would doom it to cancellation or at least a four-year wait to the 2007 window.

Spirit was slowly eased out of the PHSF building at Cape Canaveral in the early hours of 27 May, attached to the bulbous shape of the rocket's third stage. It would be taken by trailer, wrapped in a shroud and complete with flashing lights, up the road to Launch Complex 17 and the main stages of the Delta launch vehicle, waiting passively in the night for its payload. The Mission Readiness Review was held and everything appeared ready, except for the test results that would confirm whether or not the rovers could operate with a blown fuse. On 3 June, just five days prior to launch, project staff gathered in Washington DC to get the verdict from Ed Weiler and his staff. Experience with past science missions – flights where almost nothing is perfect and where the gut feeling transcends clinical logic – carried the day and the verdict was 'go' for Spirit and Opportunity!

But that was not the end of the delays, as thunderstorms blew through the Cape and held the launch for two days before Spirit finally got under way on 10 June 2003 at 1:59pm local time. However, just three days before the planned launch of Opportunity on 25 June a

BELOW In four months during late 2007, Opportunity took a series of images inside Dick Bay in the western portion of Victoria Crater, the far side being about 2,500ft away. The 20ft-tall promontory to the left is Cape Verde, while the 50ft one to the right is Cabo Frio. *(JPL)*

problem with insulation on the launch vehicle delayed the flight. Then a problem showed up with a faulty battery in the rocket's destruct system, designed to blow it up should the launch vehicle spiral out of control and threaten to crash on inhabited areas. Boeing technicians quickly 'borrowed' a good battery from another Delta further down the launch slots. But by now the Mars window was coming to a close. Finally, Opportunity got off the pad at Launch Complex 17B on 7 July 2003 at 8:18pm local time, just eight days before the close of the 2003 window.

The two MER spacecraft were not alone in their journey to Mars. On 2 June a Russian Soyuz-FB/Fregat rocket sent into space from Baikonur Cosmodrome in Kazakhstan put the European Space Agency's Mars Express spacecraft on a path to the Red Planet. Attached to Mars Express was a lander called Beagle 2, designed to separate as it approached Mars, leaving the main body of the spacecraft to enter orbit. Plunging through the atmosphere, Beagle 2 was designed to slow down through friction and survive impact by using inflatable airbags. After landing it was to unfold circular dish-shaped covers, extend a camera and begin a limited series of scientific experiments. The project had been developed by Colin Pillinger, a planetary scientist at the Open University in the UK.

Meanwhile, Spirit and Opportunity sailed on their majestic way relatively uneventful and in good health. The first trajectory correction manoeuvre for Spirit was conducted on 20 June when the thrusters on the aeroshell fired for a steady 28 minutes along the spin axis. This was followed by a series of 264 pulses totalling 22 minutes of burn time using the thrusters firing perpendicular to the axis of

rotation, the total manoeuvre increasing speed by 47ft/sec (14.3m/sec). Opportunity made its first TCM burn on 18 July when the axial thrusters fired for 54 minutes, changing velocity by 53ft/sec (16.2m/sec) and shifting the arrival date up by 1.48 days to the intended landing on 25 January 2004.

By early August the science instruments, including the ten cameras on each rover, had been checked out just to see if there had been any damage from the violent effects of launch. Although they were completely in the dark within their lander petal enclosures, 14 images were taken by each rover to provide characteristic dark-image signatures to indicate they were undamaged. Further trajectory correction manoeuvres were conducted by Spirit on 1 August, 14 November and 27 December 2003, while similar manoeuvres by Opportunity were performed on 8 September 2003 and 17 and 22 January 2004.

So accurate was the trajectory for Spirit that when controllers looked at the flight path after the final course correction burn, it was found to be off target by about 260ft (80m)! That kind of accuracy across the distance between Earth and Mars is equal to firing a bullet from San Francisco to New York and having it hit within one millimetre of the bull's eye. Of course, during the rough ride down through the atmosphere several factors would pull and tug at the spacecraft to throw it around that carefully defined touchdown spot, and its final resting place after bouncing across the surface several times was farther away than 260ft.

As Spirit homed in and the decision was made to forego two final TCM options, scientists gathered at JPL from across the United States and several other countries where universities and laboratories had contributed people and instruments to this international venture. On two floors at JPL's Building 264 teams set up camp for separately determining the daily work for Spirit and Opportunity, which after landing would be about 6,600 miles (10,600km) apart on opposite sides of the planet. While one team had their rover resting overnight, the other would be busy working away on the other side of Mars. It was the responsibility of the principal science investigator to balance the demands from individual scientists.

Both rover teams would work through the Surface Mission Support Area (SMSA), the 'engine-room' of engineering activity, to maintain healthy rovers and to monitor the

ABOVE **A superimposed MER rover gives scale to the Cape Verde promontory at Victoria Crater.** *(JPL)*

various systems on board each. It was from the SMSA that commands would flow through to the Deep Space Network and on to the spacecraft at the surface of Mars, and into which would flow the response to calls for diagnostic reports back to Earth. The SMSA processed the requests for tasks and daily activity into coded commands and computer instructions to get the rovers working as proxy geologists at their respective sites.

Each rover team was further divided into groups of scientists with common interests or disciplines, covering atmospheric science, geology, geochemistry and mineralogy, and rocks and soil samples. These four groups were subordinate to a fifth group who would be integrating the various disciplines into long-range planning and strategy. All five would work through a Science Operations Working Group (SOWG) leader chosen for his or her ability to manage and coordinate the various interests into a focused science plan. It was the job of the principal science investigator to choreograph this motley collection of diverse factions and potentially competitive interests, bringing to a common daily mission plan activities that would be processed through the SMSA for commands to each rover in turn.

For months, throughout the long, mostly quiet, time the two spacecraft were homing in on the Red Planet, each team and group were thoroughly tested with simulations and rehearsals covering every conceivable eventuality, just in case they were confronted with an unexpected problem before or after the spacecraft reached the surface. They called these ORTS – Operational Readiness Tests – and they were conducted with full-size copies of the two rovers in the In-Situ Instruments Laboratory (ISIL) located in another part of JPL. Here, in one half of the facility separated by a thick glass wall, the light levels and dust of Mars was simulated to give engineers and technicians the advantage of simulating activities on the real planet far away, testing out procedures the real rovers would have to cope with and others unimagined.

Spirit reached the surface of Mars at 8:35pm JPL time on 3 January 2004 (04:35am UTC on 4 January) and bounced 28 times to a stop 900ft (275m) away. It came to rest little more than 6.2 miles (10km) from the centre of the designated landing ellipse, at 14.57°S by 175.48°E, facing almost due south inside Gusev Crater. The combination Spirit and lander was tilted a mere 2° from horizontal with the front of the deck only 15in (37cm) off the surface. Within a couple of hours the first direct UHF data relay had been made through NASA's Mars Odyssey orbiting the planet high above – the first time that had been tried – with Spirit reporting it was fine, the lander petals had opened and the solar arrays had been deployed.

Within two days of landing Spirit had provided 12 high-resolution images which scientists used to construct the first panoramic picture of the landing site, revealing a flat surface more indicative of lava lakes than the sedimentary deposits expected. As expected, there were fewer rocks at Gusev than at the other three lander sites, covering only 3% of the surface compared to 20% at the Sojourner and Viking 1 and 2 sites. Hills about 1.2 miles (2km) distant were pinpointed as early travel locations and on 6 January NASA named these Columbia Hills in honour of the Shuttle astronauts who had died while returning to Earth on 1 February the previous year. Before that could happen, however, Spirit had to get off the landing stage, and images showed airbag material blocking the path forward on to the surface.

It would have been possible to rotate Spirit and go off in another direction but that was a potentially difficult manoeuvre, and for a couple of days engineers tried alternatively raising the forward petal and retracting the bags. In the end it was decided to attempt the tricky manoeuvre of turning Spirit on its heels and commanding it to the surface off another petal after a clockwise rotation of 115° to face west-north-west, a compass point of 286° from north. On 9 January Spirit transmitted 180MB of data, ten times the capacity of Sojourner almost seven years earlier, and a roll-off team under Joel Krajewski got to work to plan the exact sequence of events, as they had been doing since landing.

Finally, at 08:41am UTC on 15 January, Spirit completed its drive off the landing stage and on to the surface of Mars. The sequence ended with the back of the rover 2.6ft (80cm) from the landing stage, a drive completing a total run of 10ft (3m) in 78 seconds. 'There was a great sigh of relief from me,' said JPL's Kevin Burke, lead mechanical engineer for the drive-off. 'We are now on the surface of Mars.' The first destination for Spirit was a small rock dubbed Adirondack, reached three days later through a series of short arcs, Spirit first turning through 40° over 3.1ft (95cm) before a straight drive of 6.2ft (1.9m) in four movements, stopping intermittently to take images. The drive took 30 minutes, of which only two minutes had been spent in motion.

Even this short distance was a major feat compared to the limited mobility of Sojourner, and Adirondack was the first target for the robotic arm and its suite of science instruments. But within a few days Spirit fell silent and concerns grew that the rover might be failing. Early on the morning of 21 January ground controllers sent commands to the rover and it responded that it had heard, but failed to send data through the relay with Mars Global Surveyor as requested. Later that day it failed to send data through Mars Odyssey. Sporadic and intermittent data resumed but in an unpredictable pattern seemingly compromised by a software fault. Then it refused to shut itself down during the Martian night, draining power without recharging batteries from an absent Sun.

By 24 January, just a few hours before Opportunity had been due to reach Mars, engineers discovered how to solve Spirit's problem, getting it to respond appropriately and stop itself continuously rebooting. An INT_CRIPPLED command would allow the software to build a data-flow system in the RAM and avoid passing it to the flash memory, which was suspected of preventing controllers from getting control of the rover. But that proved only part of the solution; another command, SHUTDOWN_DAMMIT, had been made up by code writers to activate in the unlikely event of everything running amok during ground testing back on Earth.

Still in the software, it proved to be the only way of getting the rover to crash completely and restart itself from scratch. But it was a dangerous option, and one activated only as a last resort. If the spacecraft went silent it might never reawaken. But it did, and activities resumed after a nail-biting session living through a potential death sentence for what was otherwise a perfectly healthy rover – all thanks to the MER software architect, Glenn Reeves, who came up with the recovery plan. Now, spectacular science lay ahead on a mission that would last far beyond the expectations of anyone on the MER team.

On time and on target, Opportunity followed the successful descent profile of Spirit and reached the Meridiani Planum at 9:05pm JPL time on 25 January 2004 (05:05 UTC 26 January) at 1.95°S by 354.47°E. It touched down in the centre of a shallow crater-like depression about 65ft (20m) across, the rover facing north-north-east. The first images sent from the rover on its landing stage revealed a surface unlike anything seen at the other four sites from which pictures had been returned, certainly more instantly exciting than the view from Spirit. With darker materials and prominent bedrock, outcrops were well within rolling distance and scientists erupted with enthusiasm at their good luck in reaching such an obviously fascinating location. It was so different to the Gusev site where Spirit resided half a planet away and it promised so much.

But not every Mars scientist was celebrating. More than a week before Spirit arrived at Mars, on the morning of 25 December 2003 the Beagle 2 lander designed and managed by Colin Pillinger failed to survive entry through the atmosphere and descent to the surface. Nobody knew what had happened after it separated from Europe's Mars Express, which went into orbit about Mars and is still operating today. But disappointment was rife, and without instrumentation to send data back to Earth the probe was on its own, and there was no way of finding out when failure occurred and why. As an experiment attached to the orbiter, from which it was released six days prior to entry, it had been a stoic effort for which there had been lacklustre support from government.

BELOW A model of the Beagle 2 spacecraft designed by British planetary scientist Dr Colin Pillinger, carried by the European Space Agency's Mars Express orbiting spacecraft that failed to arrive at the surface in working condition. Designed in the form of a series of overlapping discs, it was to unfold after landing and carry out scientific investigations powered by solar cells. *(ESA)*

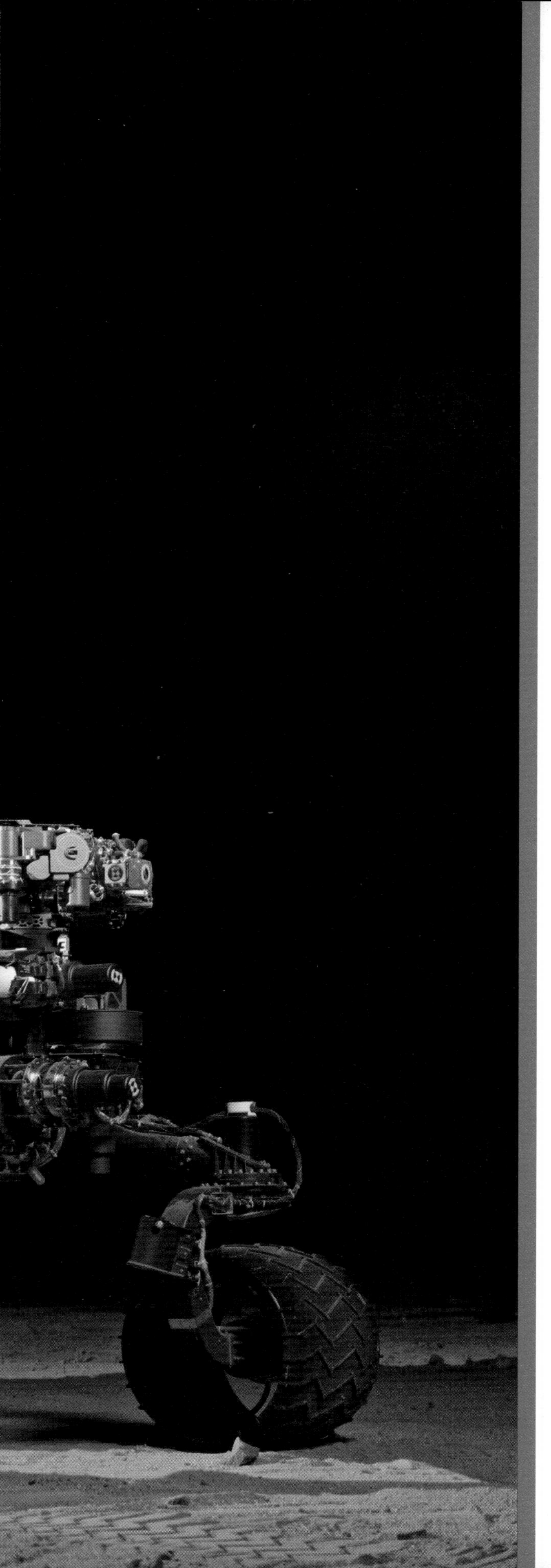

Chapter Four

Mars Science Laboratory

Challenged by demanding tasks, engineers built a new rover far larger than anything dreamed of before. Using a bold and audacious landing system, never before tested in space, and with an array of instruments and sensors that would duplicate the tool kit of human geologists, NASA set its sights on its biggest challenge yet – putting a one-tonne vehicle on the Red Planet.

OPPOSITE The Curiosity Vehicle System Test Bed (VSTB) is a duplicate of the flight hardware now on Mars, showing the cluster of contact instruments on the robotic arm. *(JPL)*

Origins

It was not known as Mars Science Laboratory (MSL) when it was conceived in 2002, but it emerged from the planning stage for a highly sophisticated rover, enabled by a search for critical technologies that began more than a year before Spirit and Opportunity were launched. It grew from a research programme that aimed to develop advanced engineering concepts into a radical method of reaching the surface of Mars. But its purpose was to provide a means of conducting in-situ measurements of surface properties and to carry out the examination of the mineralogy and the petrology of surface materials over a great distance and in a long-duration mission.

MSL was a product of the extreme demands placed upon it by the scientific objectives, which simply could not be achieved using the same technology as the MERs Spirit and Opportunity. A new way had to be found to get the rover down on the surface. It was going to be big – much larger and heavier than either MER rover. Moreover, the rover was expected to last a full Martian year – about two Earth years – so from the outset it was designed to operate for at least 500 Sols. Compared to the MER rovers' design life of 90 days it was a giant step forward.

Like Pathfinder before it, MER had used airbags to cushion the impact of landing after being dropped to the surface, but that technology could land no more than approximately 154lb (70kg) of science instruments, woefully inadequate for the complex suite of science equipment planned for MSL. Coupled with the need for a precision landing in a difficult site, a completely new method of reaching the surface was essential. It was these demanding new requirements that drove the 2002 technology review, when a bold and challenging new concept was mooted for the first time.

Instead of using a landing stage with its retro-rockets and airbags that were redundant when they reached the surface, it was proposed that MSL would use a rocket-powered sky crane hovering above the surface to lower the rover and then sever the tethers and fly away a safe distance from the site. To those for whom such exotic technologies were unimaginable, it smacked of Buck Rogers and Heath Robinson combined! Yet, with gravity little more than one-third of what it is on Earth, a one-tonne rover would weigh 380kg (840lb), thus reducing the required amount of rocket thrust needed for deceleration to a hover.

It was doable, but was it workable? NASA thought so and the sky crane concept was born, a big advantage of which was that it had development potential to put rovers on the surface weighing up to five tonnes. The MER concept was at the limit of its weight capability. Spacecraft weighing more than Spirit and Opportunity could not use airbags to survive impact. Moreover, a significant demand for greater electrical power than MER's solar cells could deliver would be essential for the wide range of scientific equipment on board. So a nuclear power source was selected to give MSL higher power and long life unchallenged by hazy skies, low Sun angles or dust on the solar-cell panels.

The success of Spirit and Opportunity after they landed in early 2004 boosted confidence in high-tech and innovative concepts using basic and proven engineering solutions. So it was with MSL, based on an evolutionary development of technology but with a radical new way of achieving the desired results. Technically, the starting date for MSL was September 2003, with the Mission Concept Review a month later on 28 October. The Formulation and Design period began in November 2003 and lasted

BELOW Just a year after Spirit and Opportunity had been conceived, in 2001 small groups at JPL began preliminary designs for a large rover to conduct a wide range of scientific studies at a selected landing site. *(JPL)*

ABOVE Mars Reconnaissance Orbiter (MRO) was launched on 12 August 2005 and arrived at Mars on 10 March 2006, by far the biggest and most advanced orbiter yet sent to the Red Planet. *(JPL)*

until August 2006, when the Development Phase began. This would last until operations commenced beginning with launch, which was planned for the September–October 2009 launch window, arriving at Mars in June 2010.

As work on MSL progressed it began to run into technical problems that forced up the budget and threatened to delay the launch. As work on MSL built up during the Development Phase, in 2007 JPL changed the heat shield material from SLA-561 to PICA (see later), and the weight of the rover grew, which in turn increased the amount of propellant the sky crane would need to hover above the surface. In June 2007 the Critical Design Review flagged serious concerns about a wide range of problems both present and looming, while NASA Headquarters put a special 'watch' tag on the entire programme. A second review in February 2008 decided that the mounting problems could not realistically meet the 2009 launch date, but a decision was held off until another major programme review conducted at NASA Headquarters in November 2008.

Managers and senior NASA officials were happy with the science equipment being developed and judged it adequate for the science goals sought, but they were unconvinced that the technical challenges of getting the rover successfully on the surface and operating for two full Earth years was feasible in the time available. Small problems were threatening to unhinge the schedule. Actuators were proving troublesome, the avionics architecture was giving problems and integration of hardware and software was proving more troublesome than expected. Options available were for outright cancellation or continuing with the programme by putting in more money and delaying the flight for two years. The following month NASA officially announced the mission was delayed to the next launch window late in 2011. More money would be found.

Business as usual

While all this was going on, Mars exploration was in its hey-day. The two rovers Spirit and Opportunity were outliving their 90-day warranty by a hefty margin and by early 2009 had clocked up five years of continuous operation. And they were not alone.

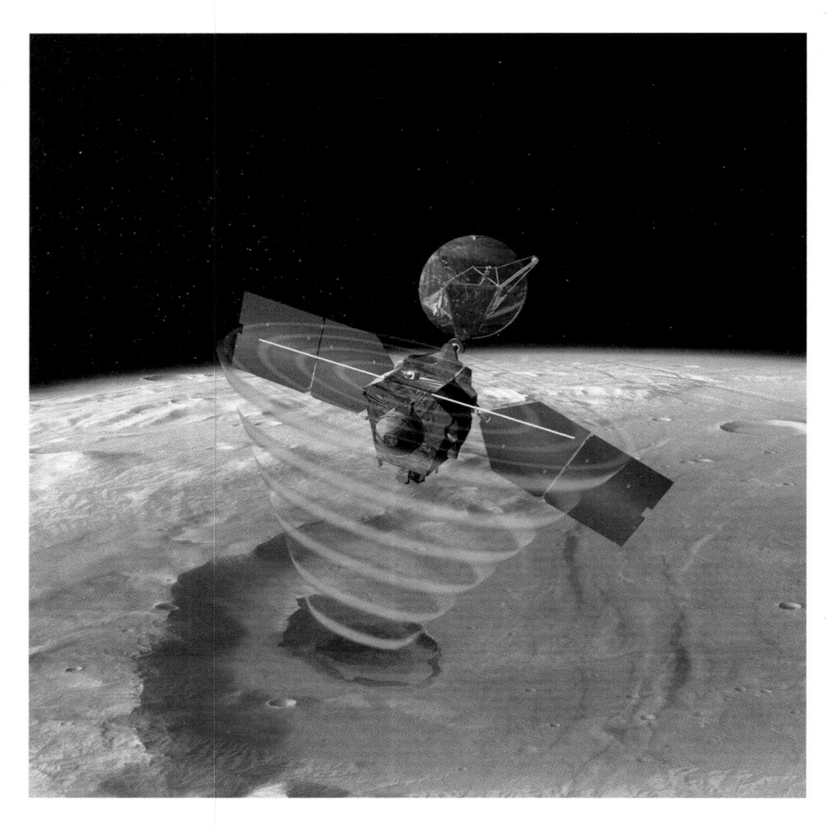

ABOVE MRO also carried a radar system capable of mapping the internal structure of ice in polar regions of Mars in a search for water that could lie beneath the surface. *(JPL)*

From orbit, as well as Mars Odyssey and Mars Global Surveyor continuing to operate, in 2005 NASA had launched Mars Reconnaissance Orbiter (MRO) and in 2008 had landed at a polar location the robot named Phoenix to sample what many scientists believed may be subsurface frozen water deposits, further confirmation of a watery evolution for the early part of Mars' 4.5-billion-year history.

Launched in August 2005, MRO was a 4,800lb (2,180kg) orbiter built by Lockheed Martin, packed with science instruments to map the surface of Mars through high-resolution images, study the atmosphere, weather systems and climate, and look for possible sites of subsurface water deposits. Like its two immediate predecessors, MRO was designed to serve as a communications relay for Spirit and Opportunity as well as the 2008 Phoenix lander.

MRO reached Mars in March 2006. After several months of aerobraking, to adjust the orbit to one only 160–196 miles (250–316km) above the surface, the initial phase of its planned science activity began in September. From an orbital plane set at 93° to the equator, the spacecraft is in a Sun-synchronous path that has it repeat an overpass at 3:00pm local time each day. On 17 November 2006, MRO conducted the first relay task when it sent signals from Spirit to Earth. With two large solar panels providing a total 2kW of electrical energy, MRO was equipped for a lengthy and highly productive life.

MRO had not long settled into its operational duties than NASA launched the Phoenix spacecraft to Mars on 4 August 2007. Again the prime contractor was Lockheed Martin and the project was managed by JPL, but this international venture was run by the Lunar and Planetary Laboratory of the University of Arizona, the first NASA Mars mission led by a university. As well as scientists and teams across the US it also involved several countries in Europe, as well as Canada. The spacecraft derived its name from the projects out of which it emerged, including design elements from the cancelled Surveyor 2001 lander and the failed Mars Polar Lander (MPL) destroyed in 1999.

Phoenix had two large circular solar array panels which gave the deployed spacecraft a width of 18ft (5.5m), while the top deck was about 4.9ft (1.5m) in diameter and the height to the top of the mast carrying weather sensors was 7.2ft (2.2m). Phoenix carried a robotic arm that could extend 7.7ft (2.35m) from the base of the lander and was able to dig down 20in (50cm) below the surface to cut through

LEFT The descent and landing profile followed closely that of the unsuccessful Mars Polar Lander, whose mission Phoenix was fulfilling. *(JPL)*

crusty permafrost and obtain samples which were delivered to instruments on the top deck for analysis. A camera was attached to the arm just above the scoop to provide close-ups of the immediate area and of the floor and walls of trenches it would dig. Eight separate one-use ovens, each no bigger than a pencil eraser, performed the combined function of a high-temperature furnace and a mass spectrometer. From this data scientists could measure water vapour, carbon dioxide and mineral content.

Phoenix also carried a wet chemistry laboratory containing four cells to which a sample was provided by the scoop and to which water was added, the products being measured by 26 chemical sensors and a temperature sensor. A thermal and electrical conductivity probe would measure thermal dissipation, thermal conductivity, electrical conductivity, wind speed and temperature. The meteorological station carried a wind indicator, and pressure and temperature sensors to send daily weather reports from the frigid surface as well as a device for measuring the number of dust particles in the atmosphere.

Built by the University of Arizona with the Max Planck Institute for Solar Systems Research, the surface stereo imaging camera was a development of the type used on Pathfinder and on the unsuccessful Mars Polar Lander and would document the surface features as they changed over time. One camera carried by Phoenix was never used, intentionally. At 904lb (410kg), including 130lb (59kg) of science instruments, Phoenix was specifically designed to study the harsh environment of the polar environment and to answer questions about potential habitability for primitive life in such a hostile region. A rover would have been superfluous in an area with flat unchanging terrain.

The method adopted for getting Phoenix down to the surface was more reminiscent of the Viking landers than that used for Spirit and Opportunity, a combination of atmospheric braking, parachutes and retro-rockets. It landed successfully on 25 May 2008 at 68.22°N by 234.25°E, some 2.55 miles (4.1km) further downrange and 3.7 miles (5.9km) to the north of its predicted track, but well within the target ellipse of 68 x 12 miles (110 x 20km). It had landed on a flat surface with a tilt of only 0.3°. It was spring in the northern hemisphere and the solar arrays would be illuminated all day and

ABOVE Phoenix carried a camera, a weather mast and a robotic arm with a scoop for collecting soil tested inside the lander. It also had a LIDAR (Light Detection And Radar) device for measuring dust and moisture in the atmosphere, represented here by the green beam. *(JPL)*

RIGHT Defined by the blue line, the area available to the robotic arm provided useful samples, localised areas receiving a variety of informal names. *(JPL)*

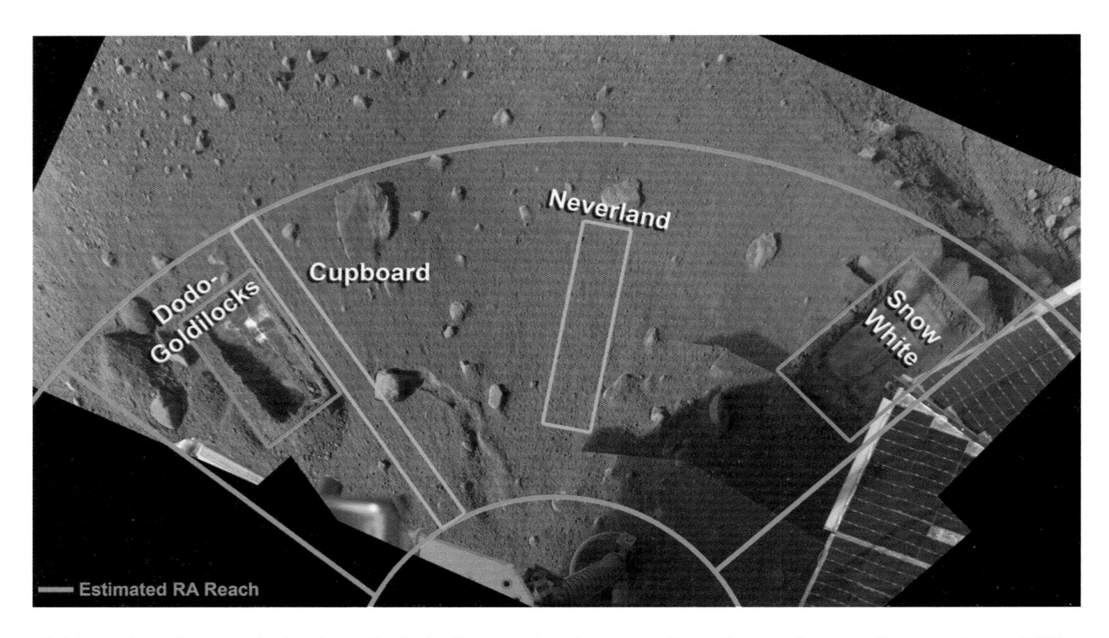

night – a location equivalent on Earth to Lapland in northern Norway – and Mars's arctic daylight would maintain a constant solar source for electrical energy. The robotic arm was extended three days after landing and this was followed on Sol 6 with its first contact at the surface, scooping up its first sample several days later.

Phoenix was never planned as a long-duration mission because it was not expected to survive the extremes of a Martian arctic environment for an extended period as the seasons changed. Temperatures on Mars can plunge as low as –256°F (–160°C), far below the survival temperature of the spacecraft's delicate components. Moreover, ice accumulating on the solar panels would crack them under weight and without them Phoenix would have no power. Designed to endure for 90 Sols, Phoenix nevertheless lasted for 157 Sols before finally it succumbed on 2 November 2008, after returning a wealth of data and scientific information.

Big science for MSL

While scientists were absorbing data from Phoenix, and continuing to reap rich results from the three orbiters around Mars, others at NASA were preparing to send the MSL rover to the surface. But this was not without its challenges. Some of the problems with development of MSL were due to the big leap forward in landing concept, the technology to make it work and the wealth of science MSL was designed to carry out.

It was universally accepted within NASA and outside the agency that MSL was the most technologically challenging vehicle ever designed to move across the surface of another world in space. Meeting those challenges had already forced a postponement in launch to 2011. The size of a small SUV, three times the weight of the two MER rovers and carrying ten times the weight of science equipment, JPL had been hard pressed to meet tight deadlines and produce a ready solution to problems that only emerged while development and testing sought a definitive design.

The plan for MSL was to launch on a powerful Atlas V 541 series rocket during a window lasting from 25 November to 18 December 2011, with arrival at Mars in the period 6–20 August 2012. Like its precursors to the Red Planet since Pathfinder, MSL would be supported by a cruise stage between injection and arrival at Mars for communications, data transfer, attitude control and trajectory correction manoeuvres. MSL's rover would land during the summer, but the mission was designed to last one full Mars year – 669 Sols or 687 Earth days. With the outstanding success of Spirit and Opportunity, however, no one doubted that the MSL rover would last a lot longer. With a nuclear power source on board, the contaminating effects of dust and sandy particles would not threaten electrical power levels, leaving only communications equipment and mobility systems to restrict the life of the rover.

The four primary science objectives were derived from knowledge acquired by previous orbiters and landers. The first was to try to discover the nature and quantity of organic carbon compounds, to search for the chemical building blocks of life and to identify features that may have been caused by chemical processes driven by biological activity. The second was to search the area for the chemical, isotopic and mineral composition of materials at or just below the surface. The third objective was to investigate evidence for the past presence of water and to determine the cyclical changes over time to both the water and the atmosphere. The fourth was to measure radiation levels at the surface and to map galactic cosmic radiation, solar emissions and neutrons reaching the surface.

A key feature of MSL was to provide scientists with a much wider range of site options by removing constraints necessary on earlier missions. Crucial to that was a virtual pinpoint landing capability that would ensure engineers could target the lander for a much smaller area, thereby putting it down in places previously considered dangerous. Site selection workshops involving 150 scientists were held in June 2006, October 2007, September 2008 and September 2010, by which final date the choice had been narrowed from over 60 sites to just four finalists. These were Holden Crater, Mawrth Hills, Eberswalde Crater and Gale Crater, the first four being 24°–27° north or south of the equator, Gale Crater being at the latitude of only 4.4° south. But the flexibility afforded by the new entry, descent and landing concept meant that the final landing site could be decided some time after launch if necessary, the precise alignment of the trajectory being adjusted en route.

Because this was the first use of the sky crane concept, engineers wanted as dense an atmosphere as they could for the parachute to bite on, providing the best achievable retardation, and while the first three sites were 3,280–7,380ft (1,450–2,250m) below the average radius of Mars, the Gale site was 14,600ft (4,451m) below the mean, thereby favouring that site for optimum engineering reasons. But the final decision, within tolerances, would be the judgement of the scientists. Steve Squyres had been project scientist on the MER rovers; John Grotzinger would carry out that role for MSL.

When selected in July 2011, the chosen site for Curiosity was indeed Gale Crater, an impact event 96 miles (154km) in diameter situated 137.4°E at a low elevation compared to the average radius of the planet. Curiosity was sent

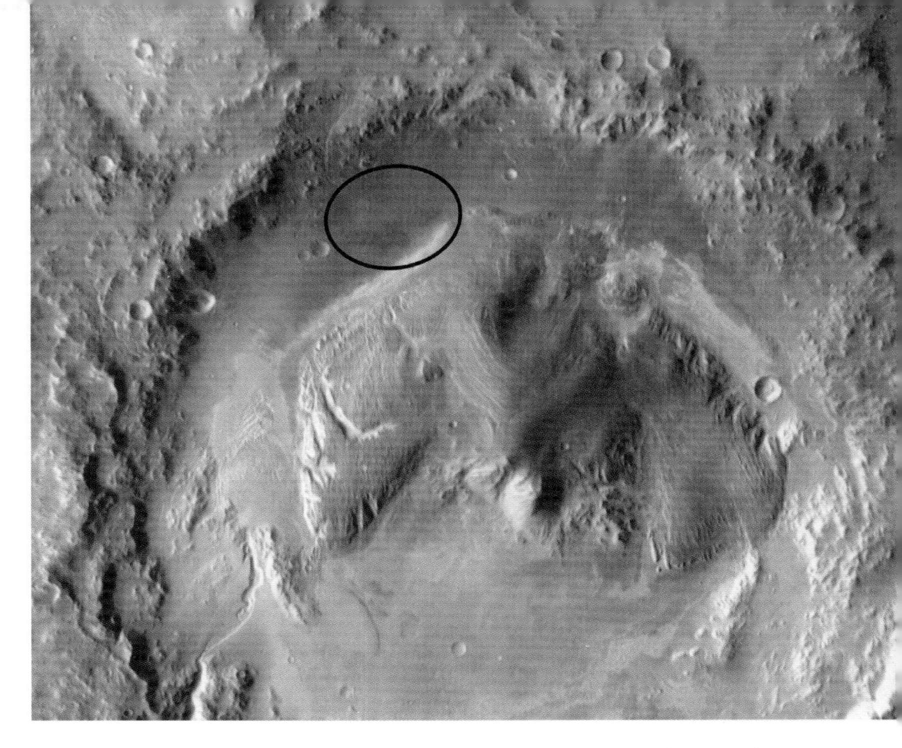

ABOVE Gale Crater had the advantage of an accessible central mountain of layered materials rising more than 15,000ft above the surface, known as Mount Sharp. Apparently deposited over many millions of years, it comprises clays at the bottom superimposed with sulphur and oxygen-bearing minerals above. *(JPL)*

BELOW MSL project scientist John Grotzinger explains the design features of the new rover while camera and imaging guru Michael Malin, of Malin Space Systems, looks on. *(NASA)*

to characterise the environment to see if it could ever have been conducive to the emergence of life – however primitive – and Gale Crater was a superb example of where those conditions might have prevailed. Because it was at so low a level, and on the edge of a much higher surface area, it was a likely place for water to have flowed freely and pooled inside the crater walls. Gale has a central mound, named Mount Sharp, rising about 16,000ft (4,900m) above the crater floor. With extensive stratigraphy, Mount Sharp had all the appearance of having built up through successive epochs of deposits during the three billion years since the crater formed. With gentle sloping foothills, Curiosity could spend many years conducting a geological and atmospheric survey around the area.

The precision promised by the Entry, Descent and Landing technique selected for MSL allowed Curiosity to reach its target with greater accuracy than any previous mission to Mars. Initially, flight controllers from the EDL team gave the mission a 99% probability of putting Curiosity within a 12.4 mile (20km) by 15.5 mile (25km) ellipse close to the foothills of Mount Sharp. In fact, as the mission progressed they shrank that to an area 4 miles (7km) by 12 miles (20km). From there the rover could drive to a wide variety of rocks, surface conditions and features, temptingly exciting in the images transmitted by orbiters around Mars.

Curiosity to Mars

Compared to previous Mars missions, MSL was a veritable giant. Curiosity was encased by the sky crane structure, which was itself inside a protective aeroshell. The sterilised aeroshell and its encapsulated spacecraft was attached to the cruise stage and the entire assembly stacked on top of the Atlas V 541 rocket situated on Launch Complex 41 at Cape Canaveral, Florida. Total dimensions of the assembled spacecraft on its launcher was a diameter of 14ft 9in (4.4m) and a height of 9ft 8in (3m) with a mass of 8,463lb (3,893kg). Of that, 1,188lb (539kg) comprised the cruise stage, 5,293lb (2,401kg) was the sky crane system and its rocket propellant and 1,982lb (899kg) was Curiosity.

The massive Atlas V 541 launch vehicle

consisted of the core stage powered by a Russian-built RD-180 engine and four solid propellant strap-on boosters with a powerful Centaur upper stage to which the payloads and the protective shroud were attached. Fully assembled the vehicle topped a height of 191ft (58m) taller than the Shuttle. The launch of MSL was critically timed so as not to compromise the brief launch window of a mission to Jupiter, set to take place from the same launch site in the period 5–26 August 2011, and still allow sufficient time to erect the MSL rocket.

At lift-off the core first stage and the four strap-on boosters produced a combined thrust of about 2.07 million lb (5,160kN), almost six times the power at lift-off of the rocket that launched Mariner 4 to Mars in 1964. The four boosters are jettisoned in pairs at 1min 51sec, followed by separation of the payload shroud protecting the cruise stage and aeroshell from the atmosphere. The first stage itself continues to fire until around 4min 37sec, when it too is jettisoned. The Centaur upper stage would then ignite and burn for about 6min 52sec, producing a thrust of 22,300lb (99.2kN), at which point the stage and its attached payload is in a low Earth orbit.

After coasting around the Earth for little more than 19 minutes, the stage reignites and fires for a further 8 minutes, achieving a speed of 22,866mph (10.22km/sec), placing the stack on a fast Type I trajectory heading for Mars. Separation of the spacecraft (the combined cruise stage and aeroshell) comes just under 43 minutes after lift-off, following which a manoeuvre by the Centaur stage places it on a different trajectory to the spacecraft. The spacecraft, meanwhile, is stabilised at a spin rate of 2rpm during the long flight to Mars.

The cruise stage itself provides communications and course-correction propulsion but gains its commands from the computer in Curiosity, still within the aeroshell. Fabricated from aluminium, it takes the form of a compressed cylinder, or ring, the diameter being approximately five times the height. A series of internal ribs is used as the mounting point for systems. The stage has a propulsion system comprising two clusters of thrusters, each with four rocket motors providing a thrust of 1.1lb (5N) of force using hydrazine propellant. Essentially a compound of nitrogen and hydrogen, the two chemicals explosively decompress when exposed to a catalyst inside each thruster. Two 19in (48cm) diameter spherical tanks contain the propellant for maintaining spin rate, adjusting the attitude orientation of the assembly and performing trajectory correction manoeuvres.

A single star scanner and two Sun sensors provide information for the spacecraft to maintain proper attitude with respect to the Sun line, and from these information is passed to the Curiosity computer to command tweaking burns with the thrusters when necessary. Electrical power was provided from six solar-cell panels with a total surface area of 138ft^2 (12.8m^2) situated around the circumference of the cruise stage. The Gallium Indium Phosphorus/Gallium Arsenide/Germanium cells would produce 2.5kW at the distance of the Earth from the Sun, but even at the distance of Mars they still provided 1.08kW even with a slant angle of 43° to the Sun. However, the radioisotope thermoelectric generator in Curiosity was able to supplement this power if necessary and if demanded by the processor.

Thermal stabilisation throughout the cruise stage and within the aeroshell was maintained by ten radiators mounted in the form of a ring around the outer edge. Coolant fluid pumped through the spacecraft's circulatory system carried excess heat from Curiosity, including that from the RTG, to warm the electronics of

OPPOSITE Standing taller than the Shuttle, the Atlas V 541 carries a Centaur upper stage that will first place the MSL payload in Earth orbit and then reignite to push MSL toward Mars. *(KSC)*

BELOW Essentially a frame supporting six solar-cell panels on the undersurface and ten radiator panels around the circumference, the cruise stage was an open structure with thermal insulation on all elements. *(KSC)*

ABOVE **Six solar-cell panel segments can be seen as light tan in colour with the radiator panels around the edge, the two clusters of four thrusters carried on tripods and seen here with red 'remove before flight' covers.** *(KSC)*

the cruise stage before excess was routed to the radiators for dispersion.

Telecommunications from launch through to just before entry was provided by medium-gain and low-gain antennas bore-sighted with the spacecraft –Z axis, in the direction of the Earth/Sun line. Because of the slightly offset centre of mass there would be a small wobble as the spacecraft spun, invoking a worst-case gain about once every 15 seconds. The job of the cruise stage was done when it separated from the aeroshell ten minutes before entry.

Entry, Descent and Landing

The aeroshell concept was essentially derived from the Viking days, the lower half known as the heat shield, the upper half being the backshell. The basic structure forms a 70° bi-conic shell with a bluntness radius half the vehicle's base diameter. At 14.8ft (4.5m) across it was one-third bigger than the Viking aeroshell and had a marginally greater diameter than the Apollo spacecraft. The aeroshell had a total height of 9.5ft (2.9m) to the top of the truncated parachute cone. It had a 0.24 lift-to-drag ratio, higher than any previously flown into the atmosphere of Mars and close to the hypersonic L/D (lift over drag) of 0.3 for the Apollo Command Module. Viking had flown a lifting, unguided entry path with an L/D of 0.18, whereas Pathfinder and the two MER rovers had flown a non-lifting entry with zero L/D.

As the spacecraft approached Mars and Entry Interface, the final Entry, Descent and Landing (EDL) updates were given to the computer at E–2 hrs with MSL 14,690 miles (23,500km) from the planet travelling at a relative velocity of 8,940mph (3,996m/sec). At just 1hr 24min prior to entry its path crossed the orbit of Deimos, a potato-shaped moon orbiting Mars at a near-circular distance of about 14,600 miles (23,460km). Some 51 minutes later it

RIGHT **With a fuelled weight of 1,188lb, the cruise stage is 13ft in diameter, almost 2ft less than the aeroshell and heat shield, and supports two spherical propellant tanks for the attitude control system.** *(KSC)*

swept past the orbit of Phobos, Mars's second and larger moon about 25 miles (40km) across in a near circular orbit with a mean distance of 9,520 miles (15,230km). From this point the spacecraft was committed to entry.

At E–10 minutes the cruise stage is jettisoned and the aeroshell is on its own as it heads towards the atmosphere. Five minutes later the flight control system issues the command to execute the EDL mode and within a couple of minutes the cruise stage would have jettisoned fluid from the heat coolant loop. Around this time too, the set of instruments and sensors built in to the aeroshell to measure atmospheric properties and the performance of the heat shield are switched on. Known as the MEDLI (MSL EDL Instrumentation), they were a vital part for monitoring the performance of the heat shield and would verify its performance. If the shield failed, engineers would get data on its performance for as long as information was sent to the appropriate orbiter for relay later to Earth.

But even if the heat shield did not fail, it would provide valuable data on how it behaved in the uncertainties of the Martian atmosphere. Bigger than any heat shield previously sent into the atmosphere of the Red Planet, it protected the combined mass of the aeroshell, the Powered Descent Vehicle and Curiosity, weighing a total 7,275lb (3,300kg), from fiery destruction. The challenge for engineers in designing a heat shield appropriate for Mars Science Laboratory as well as validating the concept and the materials for future applications at Mars and elsewhere, would benefit greatly from information on its performance. During the Apollo programme, heat shield measurements could be made after landing. This would not be the case at Mars. To provide data that engineers could use, the measurements had to be made real-time as the shield was experiencing the environment for which it was designed.

The MEDLI comprised sensors, instruments and equipment supplied by two NASA field centres: Langley Research Center in Virginia and Ames Research Center in California. Both had been involved in planetary missions before MSL. The suite of sensors included seven integrated sensor plugs (MISPs) and seven Mars entry atmosphere data system (MEADS) sensors for measuring pressure, all fitted flush with the mould line during manufacture. Each MISP sensor plug consisted of four type-K thermocouples installed at depths of 0.1, 0.2, 0.45 and 0.7in (0.25, 0.5, 1.14 and 1.78cm) from the outer mould line of the shield, providing information on the thermal profile during entry.

Each MISP plug also had a hollow aerothermal ablation and temperature (HEAT) sensor to measure a single isotherm (about

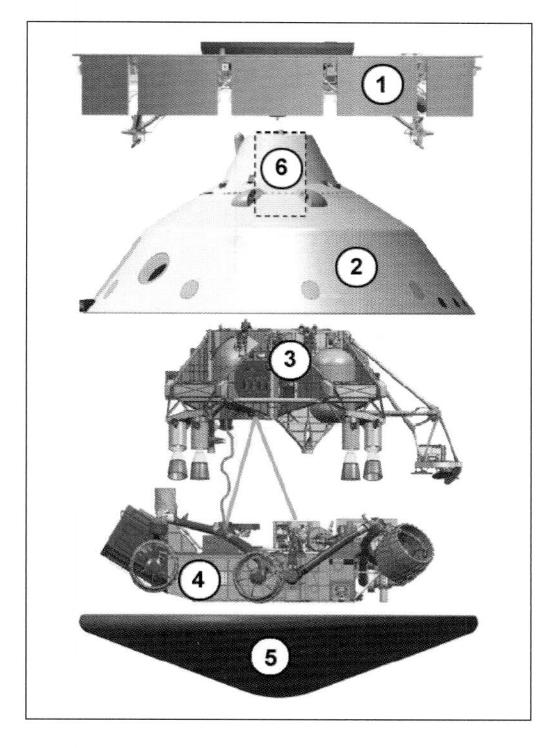

LEFT The five primary system elements include: cruise stage (1); aeroshell (2); Powered Descent Vehicle (3); Curiosity rover (4); and the heat shield (5). *(David Baker)*

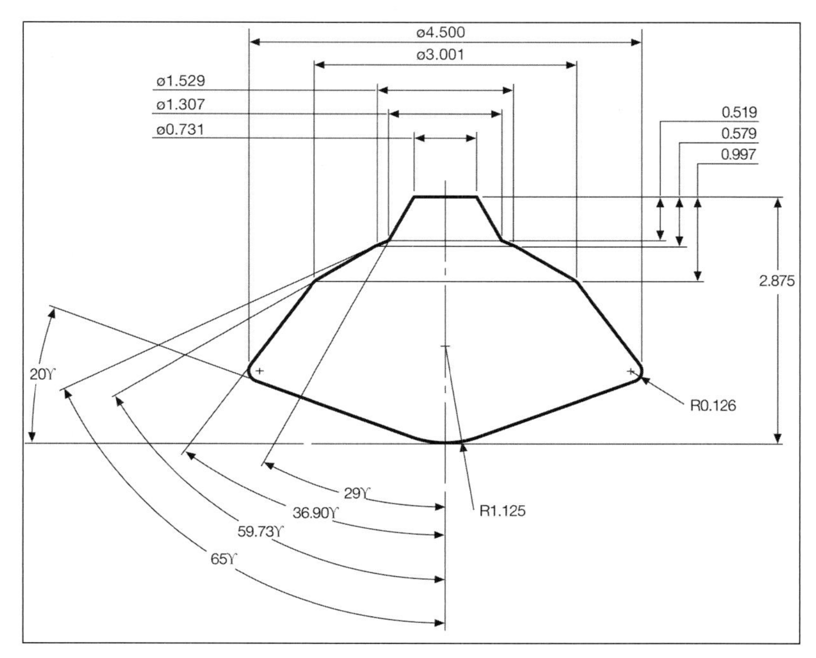

BELOW The aeroshell evolved from the Viking shell of the early 1970s and would lend itself to adaptation of Apollo guidance models for entry and atmospheric flight profiles. *(David Baker)*

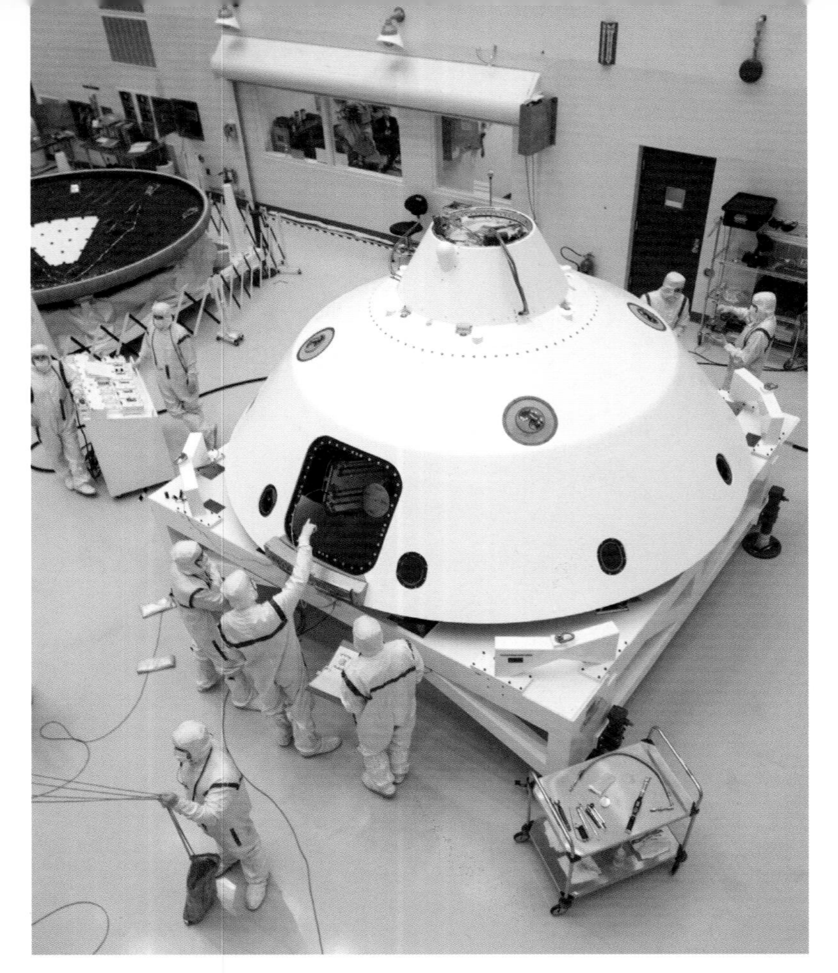

ABOVE The aeroshell has a door in the side through which the radioisotope thermoelectric generator will be inserted for installation into Curiosity while the assembled vehicle is on the launch pad. *(JPL)*

700°C) as it propagated through the shield material. This allows engineers to measure the rate of ablation at different levels over time, adding heat rate measurement to measured heat load. Embedded within the shield was the sensor support electronics (SSE) unit which provided electrical power to the sensors, signal conditioning and processing of the data for transmission as a telemetry node to the orbiters above Mars. The plugs were 1.3in (3.3cm) in diameter and were connected by wires to the SSE through holes in the aeroshell. The plugs were similar to plugs extracted from heat shields returning spacecraft to Earth, sometimes encased in plastic and given away as souvenirs after the engineers had analysed each one!

Each MEADS sensor incorporated a pressure sensor attached to the inner face of the heat shield, with a 0.6in (0.25cm) hole through to the outer surface for pressure readings. This provided engineers with flow maps as well as aid in determining the exact angle of attack at various stages of the descent. The locations of the seven pressure sensors form a cross pattern in the low-temperature, high-pressure region of the heat shield. The first data is acquired at E–10min, and continues to run at 8 cycles/sec until the parachute is deployed, but only sample data is transmitted real-time. The full data set is stored on the rover computer (Rover Computing Element) and is only transmitted after safe arrival on the surface.

At E–9 minutes Curiosity's guidance, navigation and control system (GN&CS) is turned on, which commands a sequence of manoeuvres to properly orientate the aeroshell for entry. These attitude excursions would last 3min 6sec, by which time the aeroshell was in the correct orientation with an angle of attack of approximately 18° and the 2rpm spin rate imparted to the spacecraft after launch has been nulled to zero. At E–8 minutes two solid tungsten weights, each with a mass of 165lb (75kg), are jettisoned to give the aeroshell an offset centre of mass. From E–5min 19sec the spacecraft awaits entry with reaction control system thrusters acting to null transient attitude shift through one of two GN&CS computers in the descent stage.

The dynamic characteristics of the backshell were crucial to performance. While Viking had been the only spacecraft to enter the atmosphere of Mars for a guided flight path, MSL was the first to manoeuvre on its way down. Through a series of bank manoeuvres controlling the lift vector, the aeroshell would balance energy against altitude to achieve a controlled descent to the point where it was released and the parachute deployed. The backshell consisted of a collection of truncated cones, but with several components not always flush with the smooth surface and more complex than a system of spherical, conical and toroidal surfaces.

The extreme demands of landing accuracy and greater weight to control through all phases of Entry, Descent and Landing made great demands on the design teams. Guidance would commence from the moment of entry into the atmosphere, which would require a lifting, actively controlled descent. Control of the space vehicle in pitch, roll and yaw was provided through four pairs of small rocket motors, each with a thrust of 65lb (290N), situated in the upper part of the backshell and spaced at 90° intervals. The jets were canted for directional control of the aeroshell and scarfed, with

LEFT Designed to fall away after entry, the heat shield is attached directly to the aeroshell with the rover tucked up above. *(KSC)*

BELOW Fabricated by Lockheed Martin, the heat shield is covered with a PICA material while the backshell employs the familiar SLA-561 material used for the MER rovers. *(JPL)*

the nozzle exits flush with the backshell. The reaction control thrusters would be used for all attitude manoeuvring until backshell separation.

The Entry Terminal Point Controller (ETPC), about which more later, was constructed from the Apollo Command Module entry guidance, an algorithm which had been selected for the cancelled Mars 2001 lander. The Apollo guidance programme had been written for a vehicle not so very different in size to the MSL aeroshell, which had been human-rated for its reliability and integrity. However, much work was done to modify the algorithm and adapt it to the needs of the MSL mission.

The primary purpose of the aeroshell was to protect the spacecraft from extreme temperatures from kinetic energy produced through friction with the atmosphere. During entry, temperatures could reach as high as 3,800°F (2,100°C), to withstand which the heat shield was covered with 49 separate slabs of phenolic impregnated carbon ablator (PICA), invented by scientists at NASA's Ames Research Center. This was different to the SLA-561 employed for the MER aeroshells and was made necessary by the higher mass of the

ABOVE The performance of the heat shield during entry would be monitored by sensors strategically placed at various locations. *(JPL)*

spacecraft and by the need for the structure to fly a steering entry to move the flight path closer to the desired target. SLA-561 was, however, used to cover the backshell, which would see less severe temperatures than had been experienced on all previous entry vehicles in which it had been the primary heatshield material. PICA was first flown on NASA's Stardust Sample Return Capsule, which on 15 January 2006 brought back to Earth dust samples scooped from the vicinity of Comet Wild 2.

These guided deceleration manoeuvres performed by the MSL aeroshell could be quite large depending upon the unpredictable variability of conditions during entry. Atmospheric variations including density and temperature could greatly influence the trajectory, controlled autonomously from the computer aboard Curiosity. Vital for the guided entry was the offset centre of mass, effected through the two 165lb (75kg) detachable tungsten weights carried on the backshell. With these having been jettisoned eight minutes before entry the centre of gravity shifted off-centre to allow the aeroshell to 'fly' a steerable path. The offset mass allowed the aeroshell to retain a stabilised angle to its direction of travel and to vary this angle through thruster firings, thus using variations in lift coefficient to 'fly' uprange or downrange and to remove cross-track (left or right) errors.

The purpose of the descent phase was to deliver the aeroshell to within 6 miles (10km) of the planned point above the surface for deployment of the parachute and to use the lifting entry to get there. The navigation system used a passive inertial measurement unit (IMU) rather than an active device and relied on state estimates within these limits. Because of the complexity of the shape of the aeroshell and the need for it to significantly change attitude during descent, the challenges in modelling the flight profile were far greater than had been the case with previous spacecraft. Knowledge of hypersonic gas flow and turbulence flow in the atmosphere of Mars was crucial to the final design of the aeroshell and in writing control logic software for the guidance system. Areas of atmospheric physics were uncertain at best.

The ETCP guidance segment was divided into three separate phases: Pre-bank, Range Control and Heading Alignment. The Pre-bank phase began approximately nine minutes prior to entry (E–9min), where the aeroshell has separated from the cruise stage and has manoeuvred to an angle of attack similar to the expected trim

RIGHT A MISP sensor plug of the type fitted to monitor temperature levels and heat loads during entry. *(JPL)*

ABOVE The heat shield was covered with a thermal insulation and retaining blanket. Note the sensor installations and data leads. *(JPL)*

angle. The commanded bank angle is constant at the Pre-bank value estimated to be the error it would have to work with during the more dynamic events to follow. At entry interface, 2,188 miles (3,522.2km) from the centre of Mars, the spacecraft would be at a height of 78 miles (125km), 425 miles (680km) uprange of the landing site, travelling at about 13,650mph (6,100m/sec) at a flight path angle of –15.5°.

The Range Control phase begins at about E+51sec and signals the start of guided entry. It is the point at which the aeroshell reaches

FAR LEFT In the initial phase of deployment at a supersonic speed, the parachute must unfurl in a systematic and predictive manner so as not to exert too strong a force on the lines. *(JPL)*

LEFT Fully deployed, the parachute had a diameter of 51ft supported by 80 suspension lines 165ft long. *(JPL)*

RIGHT The Powered Descent Vehicle (PDV) provides the skycrane facility for suspending Curiosity as it slowly descends to the surface. Note the four pairs of rocket motors at the cantilevered extremities of the skeletal structure. *(JPL)*

a height of 39 miles (63km), range-to-go of 250 miles (400km), travelling at Mach 29 and a relative velocity of 13,200mph (5,900m/sec). It is triggered by the sensors detecting a deceleration rate of 1.96m/s^2 (0.2 Earth g), when the guidance and navigation computer has recognised that the spacecraft has reached the 'sensible' atmosphere of Mars where it can now commence guidance control. It is here that the computer will begin to calculate the downrange distance flown and will command a bank angle to correct any errors – just as the Apollo Command Module did when returning from the Moon.

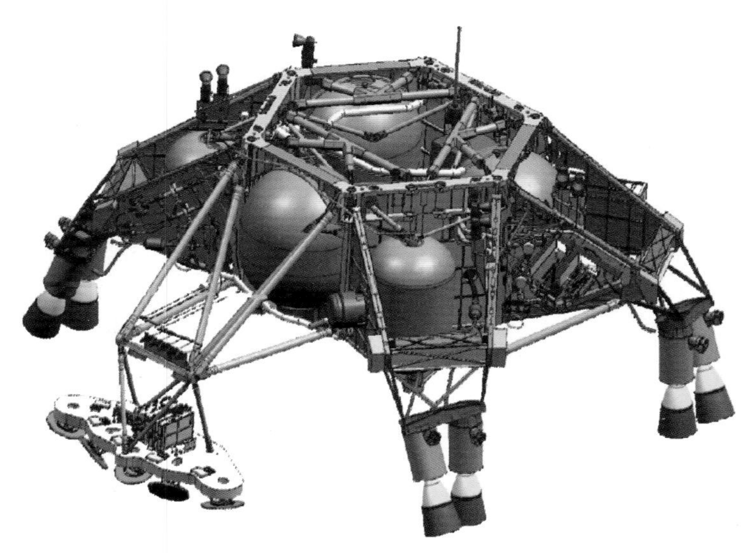

BELOW The basic frame of the PDV is supported on a six-sided structure from which the rocket motor support struts emanate. Propellant tanks fill the upper void, with the Terminal Descent Sensor, or landing radar, off to one side. *(JPL)*

This phase would see the peak heating of 165W/cm^2 on the heat shield at E+85sec. At this point the spacecraft is at a height of 17 miles (27km) with a range to go of 144 miles (230km), a speed of Mach 24, a velocity of 10,500mph (4,690m/sec) and a loading of 8g. Just six seconds later the aeroshell experiences peak loading of 11.4g, 11.9 miles (19km) above the surface but with an uprange distance of 125 miles (200km) to the landing site travelling at Mach 19 and a velocity of 8,050mph (3,600m/sec). These conditions experienced by the heat shield are well within design specifications. The PICA ablative shield is designed to withstand 216W/cm^2, 540Pa of shear and a pressure of 0.37 atmospheres.

When the estimated velocity has dropped to 2,460mph (1,100m/sec) the guidance shifts from controlling the range to starting the Heading Alignment phase. Here, the bank angle steers the aeroshell toward the target, and by limiting that bank angle to 30° the lift goes into countering the gravitational effects of the

LEFT The design arrangement of the PDV allows it to straddle the folded rover so that it presents a compact and conformal structure for encapsulation by the aeroshell. *(JPL)*

planet and increases the altitude at which the parachute is deployed. This Heading Alignment phase begins at E+150sec at an altitude of 8 miles (13km), 56 miles (90km) uprange of the landing site, at Mach 5 and a loading of 1.5g. It ends at parachute deployment 7 miles (11km) above the surface at a velocity of 900mph (405m/sec).

In short, most of the altitude is taken out during the 51 seconds of the Pre-bank phase, where very little velocity is lost, while most of the velocity is reduced in the 99 seconds of the Range Control phase, where peak dynamic loads are highest and bank reversals of up to 60° are made during the crucial energy management activity which characterises this part of the descent.

As the aeroshell slows from an entry velocity of about 30 times the speed of sound (Mach 30), the chemical reaction rates in the largely carbon dioxide atmosphere change with altitude and velocity, which have a profound effect on the dynamics of the vehicle. Predicting this and modelling the descent profile is compounded by uncertainties, which drives the requirement for the widest possible range of control logic options for the guidance computer. While the Apollo guidance equations were written for both low Earth orbit and hyperbolic return trajectories from the Moon, MSL's adaptation of Apollo's guidance logic removed the skip part of the trajectory where the spacecraft lifts in altitude before returning deeper into the atmosphere. The skip technique was written for Apollo to ensure weather avoidance should the intended landing spot be too rough for safe recovery.

Because it was desirable to get the

BELOW The rover deployment system involved a triple-bridle, data and power cable and a descent limiter to control the deployment rate as Curiosity separated from the PDV, all fed through a single confluence point. *(JPL)*

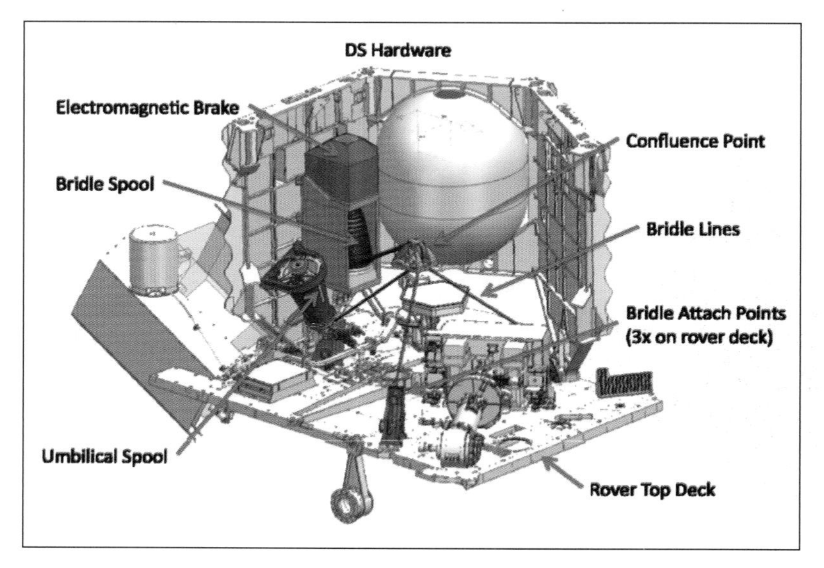

parachute deployed at the highest possible altitude, trajectory design engineers set themselves the minimum deploy altitude of 3.75 miles (6km), preferring at least 5 miles (8km) to allow sufficient time for maximum deceleration while not compromising the ETCP phases. The lower the descent rate at the time the sky crane separates from the backshell and begins powered descent, the less propellant required to come to a hover above the landing site. But the velocity at parachute deployment must not be too high for fear it will burn up through excessive heating, or tear itself apart due to a violent inflation dynamic. The design parachute opening load is 65,000lb force (289kN) within a deployment velocity spectrum of Mach 1.1–2.3. By optimising the trajectory in the ETCP phases, EDL engineers managed to lift the parachute deployment altitude to a credible 7 miles (11km)!

BELOW The bridle spool and electromagnetic brake is set up at the top of the PDV, each bridle attached to a connecting point on top of Curiosity. *(JPL)*

Acceleration and heating constraints are matched by entry angle and trajectory bandwidths, but another crucial constraint is based on the arrival time in the atmosphere, and that concerns the requirement to get real-time telemetry relayed to Earth by one of the orbiters at Mars. Since the failure of Mars Polar Lander in 1999, NASA has stipulated all spacecraft must transmit data so that engineers can examine the information and find out what went wrong should the mission fail – there is no 'black box' flight recorder to recover and no wreckage can be retrieved. It would all depend on the data streaming live to Earth, so the entry had to be synchronised with one of the Mars orbiters in direct line of sight with MSL as it entered the atmosphere.

Communication with Earth after separation of the cruise stage was effected through two X-band antenna on the backshell: the 'parachute low-gain antenna' and the 'tilted low-gain antenna'. Communication with Mars orbiters would be made through the parachute UHF antenna. The final task prior to parachute deployment was to shift the aeroshell's centre of gravity back to the axial symmetry of the spacecraft again by releasing, from E+230sec, a further six tungsten weights, each with a mass of 55lb (25kg), at the rate one every two seconds from Mach 2.5 to just before parachute deployment. The parachute itself was contained inside a conical structure on the top of the backshell, and deployment would occur at an altitude of 7 miles (11km) and a velocity of about 900mph (405m/sec) at E+254sec. The spacecraft would remain on the parachute for just under two minutes.

The parachute was of the now familiar Disc-Gap-Band type first employed on the two MER rovers. For MSL it had a diameter of 51ft (15.5m) compared to 37.7ft (11.5m) for Phoenix, 49ft (15m) for MER, 40.7ft (12.4m) for Pathfinder and 54.1ft (16.15m) for Viking. The white nylon of which the parachute was fabricated was reinforced by a stronger polyester at the vent in the apex of the canopy to resist higher stresses in that region. The canopy was supported by 80 suspension lines and the inflated assembly was 165ft (50m) long. It was designed to withstand deployment at up to Mach 2.2 and had promising development

LEFT Two important instruments looking at the surface after heat shield separation are the radar antennas and the MARDI descent imager, seen here, which takes a series of stills to compile a moving sequence of the landing phase. *(Malin Space Systems)*

potential for more advanced missions using this type of descent and landing technology.

Heat shield separation would occur 24 seconds after parachute deployment with the spacecraft heading toward the surface at an altitude of just below 5 miles (8km) and a velocity of 280mph (125m/sec). This event would be triggered by the inertial measurement unit sensing a velocity of Mach 0.7 or less. At this point the Mars Descent Imager (MARDI) would start recording high-definition video of the descent in natural colour. Looking straight down at the surface it would provide a continuous record of the approach phase for the last 25,000ft (7,620m) down to the surface. This would not be broadcast live but recorded and sent to Earth after touchdown from the orbiter – whatever the outcome.

From the time the heat shield is jettisoned to the time Curiosity reaches the surface, MARDI would take about 500 images of 1,100 x 1,200 pixels in size. The first image taken at an altitude of around 3 miles (4.7km) would have a surface resolution of 2.5m/pixel and cover an area of 4 x 3km. The last picture would be taken at a height of just 19in (0.5m) and provide a resolution of 1.5mm/pixel over an area of 38 x 20in (1 x 0.75m). When scrutinised on Earth, the descent images would allow scientists to define the exact landing spot with extreme accuracy. By measuring the degree of drift and associated compensation made by the descent landing system, MARDI images would give scientists data to plot winds down to the surface. These images would also be used to help plan traverses across the surface by showing a broader expanse of the crater floor at a higher resolution than that which would be possible from orbit.

Contracted to NASA via JPL, Malin Space Science Systems (MSSS) was selected in December 2004 to provide three camera systems for MSL, one of which was this one. Malin has carved out a secure niche as one of the world's leading exponents of high-precision science instruments for space operations, not least for the outstanding performance of its imaging equipment carried on several NASA spacecraft. MSSS provided two cameras for Mars Reconnaissance Orbiter, an earlier version of the Mars Descent Imager for the Phoenix polar lander, as well as the earlier Mars Global Surveyor.

The MARDI camera head would be mounted to the outside of the rover body connected by cable to the Digital Electronics Assembly (DEA) situated within the temperature-controlled compartment in the interior of Curiosity. MARDI shares several components with the MastCam and the MAHLI (see later) and uses a Bayer Pattern Filter CCD to ensure reproduction of natural colour to provide images at least equal to those achievable by commercial digital cameras

ABOVE The second device looking down after shield separation is the nest of radar antennas providing data for velocity and altitude measurements, seen here in the foreground covered in insulation. *(JPL)*

BELOW Autonomous control of the descent profile was the responsibility of the computer in Curiosity fed with signals from the radar antennas, tested in the air from a helicopter. *(JPL)*

on Earth. In that respect it provides images equal to those seen by the human eye. Another aspect was the possibility of stringing together all the images to produce a movie of the full descent phase from an altitude of 3 miles (4.7km).

MARDI would not be the only instrument looking at the surface after heat shield separation. Carried on the descent stage would be a pulse-Doppler velocimeter/altimeter, the Terminal Descent Sensor (TDS), which uses Ka-band multiple radar antennas to measure the decreasing distance to the surface and also both vertical and horizontal descent velocity. All events up to this point will have been controlled by altitude pressure measurements or by timed sequences. Now they would be controlled by radar measurements of the surface.

The TDS comprises six independent radar beams, three canted 20° off nadir (straight down), two canted 50° off nadir and one looking directly down. The array weighs 55lb (25kg) and measures 4.3 x 1.6 x 1.3ft (1.3 x 0.5 x 0.4m). It operates on a frequency of 20GHz with a beam width of 3° and transmitter power of 2W per beam drawing 30W of electrical power and transmitting at 120W. Each antenna has a dedicated front-end filter assembly and transmit/receive module and the

TDS cycles through the beams one by one at 20Hz (50ms intervals).

As the descent stage drifts toward the surface on its parachute, the precise altitude would vary according to the different surface elevations below. The radar would become active eight seconds after heat shield separation, at about EI+286 sec, so that it would not pick up signals from that structure and send confusing data to the guidance and navigation unit. By delaying the TDS activation so that the shield has fallen sufficiently far away – at least 56ft (17m) – the radar would acquire the ground with at least five of the six beams. The radar system measures altitude between maximum and minimum values. Should one beam acquire the heat shield and other beams acquire the surface, the radar would deliver data that indicated the mean between the two was in fact the surface. This would send conflicting commands to the Powered Descent Vehicle (PDV).

The TDS required at least five seconds of data, ideally ten seconds, after activation to get the initial set of landing-site profiling measurements. It would need to correct for slant-angle data acquisition from the offset angles of the antennas. Nevertheless, the TDS would be the only means of navigation from backshell separation down to the surface. Another function of the radar was to make sure there were no surface rocks, boulders or shelves which could breach the belly pan of the rover, nominally 24in (0.6m) above a flat surface when the rover is at rest. The radar, therefore, must position the descent stage above an area of at least 43ft^2 (4m^2) (the approximate undersurface area of Curiosity) free of obstacles no higher than 22in (0.55m).

The final sequence of events would cover the last 130 seconds of the EDL phase and would be critical to the safe landing of Curiosity. Backshell separation would occur at a descent rate of 268mph (120m/sec) and an altitude of 5,250–6,560ft (1,600–2,000m) as determined by the TDS sensors. It would cast off the aft section of the aeroshell and the parachute section and set free the Powered Descent Vehicle with Curiosity tucked up inside, its wheels folded against the sides and under the belly pan. It would be the function of the PDV to deliver the 'payload' (Curiosity) to a position from where it could be gently lowered to the surface, and then to cut the tethers and fly away a safe distance before crashing to the surface.

Key to delivery of the spacecraft to the hover point were the eight throttleable Mars Lander

ABOVE All five elements of the Mars Science Laboratory come together at NASA's Jet Propulsion Laboratory. *(JPL)*

Engines (MLEs), designed and developed by Aerojet Corporation and based on the company's highly successful MR-80 Viking Lander terminal descent engines. The primary difference lies in the single nozzle carried on the PDV compared to the 18-nozzle configuration for Viking, and in the selection of S-405 rather than Shell 405 as the catalyst due to a change in manufacturer. Both Viking and MSL nozzles have a 16.7:1 expansion ratio. Designated MR-80B for MSL, the engines have a two-layer radial outflow catalyst bed with high-purity hydrazine as propellant, the inlet supply pressure being 600–760lb/in^2 (4.14–5.24MPa).

The design requirement for MSL was more demanding that than defined for Viking, with a thrust range of 90–674lb (400–2,998N). By design adjustments the propellant flow rate was varied to a range of 0.045–3.6lb/sec (0.020–1.633kg/sec), thus providing an infinitely throttleable thrust range of 7–810lb (31–3,603N). A valve response time of 50m/sec was demonstrated, as well as a total qualified burn time of 350 seconds. A total of seven engines were used for development of the MR-10B, which incorporated a new Moog valve assembly versus the Montek E-system valve used on Viking. During the test and qualification tests, in 2007 it was decided to fit a nozzle extension, increasing the expansion ratio and narrowing the divergence of the exhaust plume.

The Powered Descent Vehicle was essentially a skeletal structure designed to hold the rover, provide a mounting position for the radar and antennas and outrigger positions for the four paired MR-10B landing engines. It carried 860lb (390kg) of monopropellant hydrazine in four separate tanks, two inertial measurement units, an electrical system comprising two non-redundant 27–37V thermal batteries, two redundant 26–36V pyrotechnic batteries and a bridle system consisting of three tethers and a single data umbilical connecting Curiosity to the PDV. Under a nominal descent profile, there would be a propellant margin of approximately 198lb (90kg).

The powered descent was divided into separate functional stages, the first of which was to follow a 3D polynomial trajectory to reduce velocity down to a descent rate of 45mph (20m/sec) before commencing the constant velocity phase. In this phase the PDV would bring the combined PDV/Curiosity assembly down to an altitude of 345ft (105m), some 985ft (300m) perpendicular to the plane of the EDL trajectory, so as to avoid landing on top of the backshell and parachute. These elements would reach the surface before Curiosity because they were not decelerating. At backshell separation the vehicle

BELOW For the first time a spacecraft entering Mars's atmosphere flies a guided and lifting trajectory. *(JPL)*

RIGHT After entry the parachute deploys and begins slowing the spacecraft from a speed of 900mph. *(JPL)*

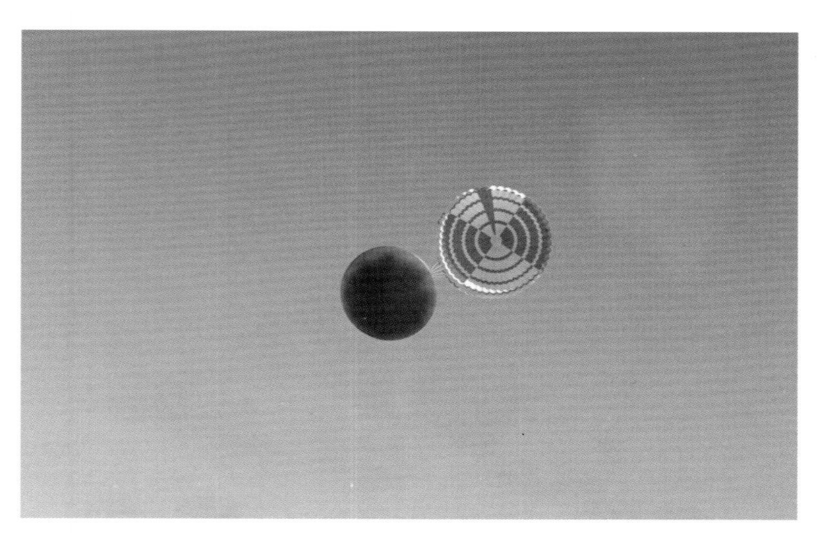

would still have some horizontal motion and this would expose the landing radar to data that was not a true measure of the precise touchdown point. Because of the motion of the PDV the fixed radar antennas could sweep the surface and incorrectly show considerable variation in altitude.

Backshell separation from the PDV would be achieved by firing pyrotechnic separation nuts, at which point the PDV would drop away, increasing speed to 280mph (125m/sec) within the next three seconds. The eight Mars Landing Engines would fire and remain at 1% thrust for 0.8sec, followed by throttle-up to 20% for 0.2sec, at which point they were warmed up and ready to take guidance commands, initiated by 2.2sec of rate damping to stabilise the vehicle. Attitude excursions would be corrected by commands to the eight engines from the two inertial measurement units.

The first phase of powered descent would begin 3.4sec after backshell separation, with the assembly at an altitude of about 5,315ft (1,620m) and falling at 280mph (125m/sec). For the next 32 seconds the powered approach

ABOVE At a speed of 280mph the heat shield is jettisoned, allowing the terminal descent radar and the descent imaging camera to see the surface below. *(JPL)*

LEFT Dropping free of the backshell and parachute assembly, the Powered Descent Vehicle fires all eight terminal descent and landing engines to decelerate from a speed of 280mph. *(JPL)*

phase would eliminate horizontal velocity and gradually reduce the descent rate to 45mph (20m/sec) by the time it had descended to a height of 345ft (105m). To this point the PDV would have consumed about 426lb (193kg) of propellant, about 49% of the total supply. This would mark the constant velocity phase, holding the same descent rate when it reached a height of 180ft (55m), some 35sec after backshell separation. Just 2.5sec later it would start the constant deceleration phase.

Here the guidance computer would follow a constant deceleration curve from 45mph down to a near hover, slowing to a mere 1.7mph (0.75m/sec) some 69ft (21m) above the surface. At this point, about 40 seconds after backshell release, four of the eight engines would be shut down to increase the degree of precision in balancing the structure in the hover. At this height and holding the very slow descent rate, the guidance, navigation and control system would shift to the sky crane mode, and about 42.5 seconds after backshell separation Curiosity would separate from the PDV by firing pyrotechnic separation bolts. From the end of the powered approach phase to the commencement of the sky crane mode, the PDV would have consumed about 99lb (45kg) of propellant, or little more than 11% of the total available.

From this point only three tethers would hold the rover, attached at the top to a bridle housing under the PDV close to the centre of mass and at the bottom of each at three dispersed places on the top deck of Curiosity. The triple nylon bridles were 25ft (7.5m) in length when fully uncoiled from a spool and fed through the Bridle and Umbilical Descent Limiter (BUDL) to which the tethers and the umbilical were attached. Immediately after rover separation the bridles would start to unwind and the front rocker on the rover would be deployed.

BELOW Its wheels still in a folded configuration, Curiosity is lowered beneath the PDV by the bridle assembly unreeling it to a length of 25ft on only four descent engines. *(JPL)*

Within three seconds, 39ft (12m) above the surface, the rover wheels would be deployed, followed two seconds later by release of the bogie, and after seven seconds Curiosity would have been lowered to the full length of the tethers, 29ft (8.8m) above the surface. The BUDL reached its limit when the brake attached to gears and mechanical resisters was applied. The resulting snatch motion would be corrected within two seconds by the guidance system differentially throttling the four engines. The last mechanical action would release the rover's differential. The touchdown logic in the computer system would begin functioning on the snatch reaction, maintaining the slow descent rate and monitoring the four MLEs at one-second intervals.

At touchdown of the 1,982lb (899kg) rover the PDV would sense surface contact due to the offloaded mass, with all six wheels on the surface within 1.7 seconds. After pausing for one second a command would be sent to cut the bridle tethers at the rover end, and the descent controller would wait for 187 milliseconds to confirm clear tether severance before throttling two engines up to 100%, the other two to a slightly lower thrust, causing the PDV to tip over and fly away on an arching trajectory of 45°. When the turn was complete the other two engines would ramp up to a full 100% thrust and, trailing its tethers and umbilical, the vehicle would continue to fuel depletion, crashing to the surface at least 490ft (150m) from Curiosity, four seconds after touchdown. The sky crane phase would have used up about 93lb (42kg) of propellant, a little under 11% of the quantity loaded at launch.

Getting Curiosity safe on the surface of Mars was fraught with potential problems and critical event sequences, each of which had to happen precisely as planned. The rover could be put down with a horizontal velocity of up

BELOW Contact is made at the surface with the Powered Descent Vehicle operating as a sky crane. *(JPL)*

RIGHT As MSL streaks in to a landing, Mars Reconnaissance Orbiter crosses the site taking images of the spacecraft as it descends in a masterful piece of synchronised operations. *(JPL)*

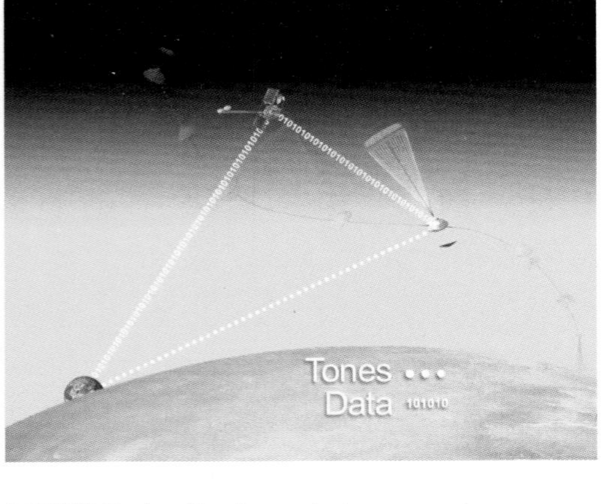

ABOVE During the descent phase a series of 'tones' signalling progressive phases of the descent would be sent from MSL to Mars Reconnaissance Orbiter, which would use its High-Gain Antenna to relay those events to Earth. *(JPL)*

ABOVE The descent profile pioneered by MSL provides a landing system with great potential for future missions involving heavy landers or rovers. *(JPL)*

BELOW Most speed and range guidance was achieved during the 50–150 second period after entering the atmosphere of Mars. *(JPL)*

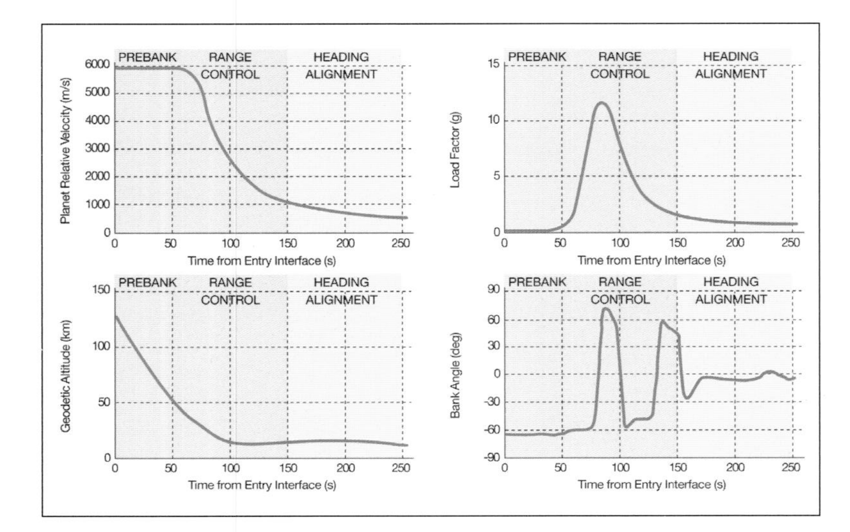

to 0.5m/sec or a vertical drop of 0.85m/sec on a slope of up to 15° and on a rock up to 0.55m in height. Ideally, except for a descent rate of 0.75m/sec at touchdown, all those parameters should be zero. The separation of various elements from just after launch from the Earth to a safe landing required 76 pyrotechnic devices to work within milliseconds of their assigned time.

Crucial to these events were the 510,000 lines of code needed to get instructions into the spacecraft assembly for the EDL phase alone and there was no available weight for redundant systems. There was only one of everything, and each had to work 100%. Moreover, while separate parts of the mission could be simulated there was no means of simulating the entire mission as it would unfold. And there was nothing other than computer codes that could access any part of the system for tinkering or making repairs.

The Curiosity spacecraft

The MSL rover was named Curiosity in a competition during late 2008 and 2009, which attracted 9,000 entries from students aged between 5 and 18 years, its name being selected in 2009 from an essay by Clara Ma of Lenexa, Kansas. She wrote that 'Curiosity is such a powerful force. Without it we wouldn't

be who we are today. Curiosity is the passion that drives us through our everyday lives. We have become explorers and scientists with our need to ask questions and to wonder.' Clara Ma went to JPL and signed the rover that now bears her name on the surface of Gale Crater.

Curiosity has a length of 10ft (3m) discounting its robotic arm, a width of 9ft (2.7m) and a height to the top of its Remote Sensing Mast (RSM) of 7ft (2.2m). With the robotic arm fully extended, it has a total length of 15.4ft (4.7m). A hefty 165lb (75kg) of its total weight of 1,982lb (899kg) is taken up with science equipment, more than nine times the 20lb (9kg) carried by each MER rover. The rover is built

ABOVE The Curiosity rover is the size of a small SUV, and a full-scale engineering model towers over the diminutive Sojourner (centre foreground), not much larger than Curiosity's toolkit at the end of its arm. It overlooks even the single representative of Spirit and Opportunity (left). *(JPL)*

RIGHT Dubbed 'Scarecrow', the mobility test vehicle displays all the critical design features of the rover's rocker-bogie system but weighs only one-third that of Curiosity to simulate the weight of the six wheels on Mars. *(JPL)*

RIGHT Routinely paced around the Jet Propulsion Laboratory's 'rock garden', Scarecrow gets a workout by testing its mobility and checking ways to avoid getting stuck in dust or overwhelmed by boulders. *(JPL)*

around its mobility system, inherited in principle from a line of development traced back to Sojourner's rocker-bogie system, its radioactive electrical power source, a set of science equipment and its robotic arm.

The mobility system is a greatly scaled-up version of the system used on Spirit and Opportunity. It has six drive wheels, each 20in (50cm) in diameter, compared to 25cm for the MER wheels and 12.7cm for the wheels on Sojourner. Curiosity has a wheelbase of 68in (172cm), a track of 87in (210cm) with the fixed centre wheels located 3.3in (8.3cm) outside the track of the front and rear wheels. The aluminium wheels have cleats for structural rigidity and curved spokes to provide some degree of flexing for absorbing loads on landing and providing a softer ride over rocky surfaces alleviating shocks on rover equipment and systems. The rocker-bogie system has a ground clearance of 26in (66cm) up to the belly pan, with the top deck of the rover 44in (112cm) above a flat surface. The geometry and scale of the mobility system with its four-outer-wheel steering is based on a requirement to turn in place and to drive in tight arcs.

As with the MER rovers, the rocker-bogie system allows all six wheels to remain in contact with the surface, irrespective of terrain. One bogie connects the middle and rear wheels on each side and provides a pivot point between them. The rocker portion each side connects the bogie to the front wheels. The drive systems incorporate an electric motor and gearbox for each wheel, geared for torque and not for speed. The drive system can tolerate a static tilt of up to 50°, with a tilt limit of 30° while driving. It has the capacity to scramble over rocks up to 20in (50cm) in height and has a top straight-line speed of about 1mph (4cm/sec) over flat hard ground, or an average 1.5cm/sec during autonomous drive with hazard avoidance. Like Opportunity and Spirit, Curiosity can dig by using one corner wheel alone, leaving the others immobilised.

BELOW The evolution of wheels on Mars, from Sojourner (1997) to Spirit and Opportunity (2004) to Curiosity (2012), shapes different configurations according to the weight of the rover and the diameter of the wheel. Titanium spring inserts on Curiosity's wheels help cushion shocks. *(JPL)*

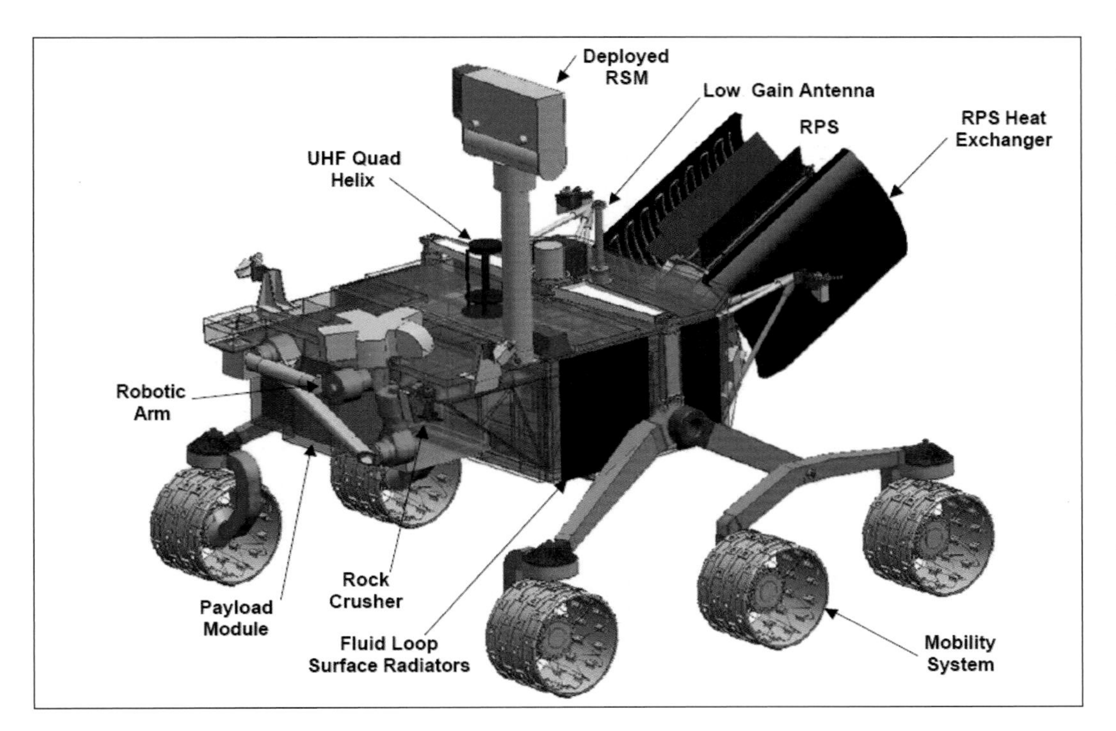

LEFT The arrangement of the rover was based upon a structural platform for mounting experiments on top and at the front slung beneath a strongback connecting the mobility system at the rocker-bogie attachment interface, which is anchored at the mid-body position. Power would come from a radioisotope electric generator carried at the rear, with a system of navigation and hazard-avoidance cameras. The main body contained the Warm Electronics Box with the computers, processors, memory and selected instruments receiving data from sensors on the robotic arm or the mast. *(JPL)*

The main body of the rover is the Warm Electronics Box (WEB), another legacy from the MER rovers, with the top equipment deck, 78 x 39in (2m x 1m) in size, providing a hard mounting for the external instruments such as the RSM and the High-Gain Antenna (HGA). The WEB also houses the Rover Compute Element, Curiosity's central computing unit, which through an industry-standard interface box provides communications and control of all the electrical motors, science instruments, cameras, navigation sensors and several other functions essential to engineering and science activity. The memory can tolerate levels of radiation anticipated for the Martian environment and is capable of protecting the rover against power cycles, outages and power system malfunctions, ensuring no data loss due to sudden losses.

The computer system incorporates two redundant Rover Compute Elements – 'A' and 'B' – with one active and the other held as a backup. Most systems and instruments can be controlled from either RCE, except for the Navcams (see 'Eyes on Mars' section, below). The active side of the computer had been the primary computer for MSL during its long trip to Mars so it had been thoroughly worked and engaged throughout. If the active side should have failed for some reason during EDL, the other side would automatically cut in and deliver commands and instructions to the entry systems, albeit to a skeleton level, operating as a 'hot' backup.

Each RCE has a dedicated radiation-hardened central processor with PowerPC 750 architecture, a BAE RAD 750. This operates at a speed of 200MHz, ten times the speed of the single RAD6000 central processor in each MER rover. Each processor communicates with peripheral devices using other cards connected on a compact peripheral component interconnect (cPCI) backplane interface, which also provides central memory storage for essential mission date and telemetry of 32 gigabits via a non-volatile NVMCAM card.

Each computer element has 2GB of flash memory, eight times that in the MER rovers, with 256MB of DRAM and 256KB of electrically erasable programmable ROM. The software in Curiosity was responsible for monitoring the health of the rover from the time it departed the launch vehicle through to initial surface operations, but thereafter it is reprogrammable with better, updated and improved software as the mission progresses and as new codes are written. The first in-mission upgrade took place in May 2012 followed by another in June that year ready for installation during the several days after landing. Redundant power and analogue modules are connected to the RCEs via MIL-STD-1553 data bus connections.

Software in the main computer will conduct health checks on the spacecraft, search for commands from Earth and run the entire flight control system from software in each RCE, an architecture legacy from the MER rovers. It will even watch for potentially incompatible activities that could be affected by radio frequency interruptions while Curiosity is communicating with Earth.

All communications information is stored in either the primary table or the high-priority table, the former holding as many as 256 communication windows which are used for standard operations on the surface. The high-priority table would be used for solving anomalies, where it will automatically cut in. Any windows loaded in the high-priority table automatically take precedence over the primary. This duality permits selected windows to be replaced without potentially compromising the originally planned communications sets.

The software would be essential for safely operating Curiosity across the Martian surface and navigating paths between obstacles and on selected routes to distant destinations. These can be achieved through three basic modes: blind-drive, hazard avoidance and visual odometry. Each or a combination of all three can be programmed into the rover for any typical day. In blind-drive mode Curiosity calculates the distance it is commanded to drive through measurements of wheel rotation – a full rotation of the wheel taking the rover almost 25in (63cm) across the surface. This is only applicable for routes where planners know there are no hazards in the way.

For more difficult traverses where images show potential hazards in the path, the

BELOW Comprising chassis and structural body, the lander is assembled with its mobility system and U-shaped wheel hangers each with its own electric motor. *(JPL)*

ABOVE The wheels with their electric motors and the arched hangers for added shock-suppression are displayed during reinstallation after tests, 28–29 June 2010. *(JPL)*

BELOW The first drive test, 23 July 2010. Sadly it will never leave the building under its own power! *(JPL)*

ABOVE The rocker-bogie arms act as travel-runs for wiring and conduits for sensor cables feeding engineering information to the onboard computer. *(JPL)*

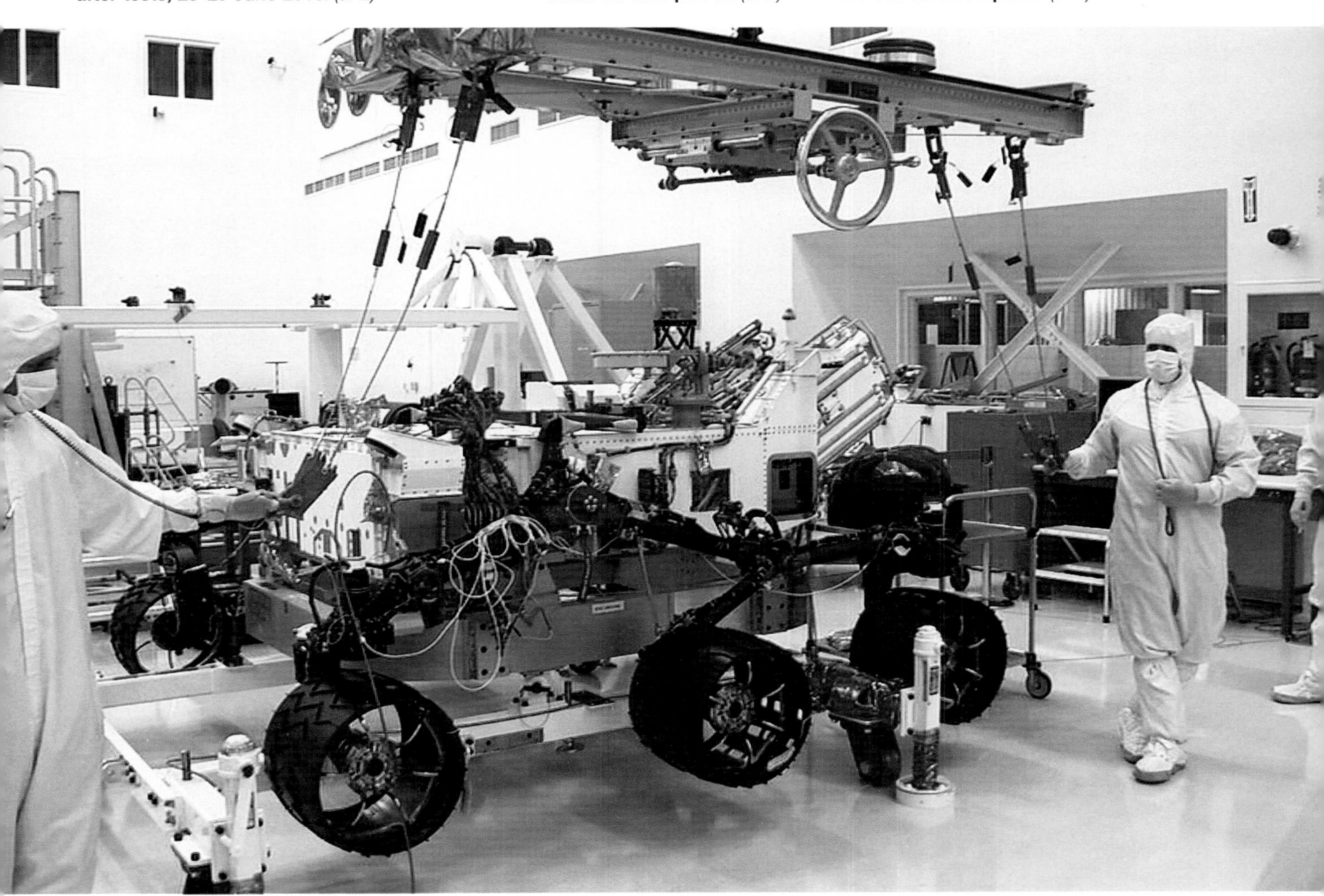

ABOVE Supported in a holding cradle, the mobility system is checked with robotic arm and remote sensor mast in place. *(JPL)*

computer controls the drive by stopping the rover at frequent intervals to take stereo images, assess the danger and decide from a pre-programmed matrix whether to stop driving or plot a safe route around the obstacle. An odometry drive requires the rover to stop frequently and take images using the Navcams to see where it is going, where it has been in the intervening segment of the drive, and what lies ahead. From this data it calculates the distance planned versus the wheel rotation rate, and any difference indicates wheel slippage. Options exist for a slip-limiter, which if reached will command the rover to stop and consult with the planners on Earth about its next action.

The computer is essential because it gives Curiosity autonomy and the essential decision-making impossible with earlier Mars vehicles. Odometry driving can extend from simplified sideways imaging to confirm calculated versus measured distance, to more complex, interactive software tools using gyroscopes in the inertial measurement unit to assess tilt and other attitude indicators of the terrain. The computer is the means by which the science instruments and tools are used and incorporated into an integrated operational plan for each day's activity. It is also the means by which the High-Gain Antenna (HGA) is kept pointing at Earth and in the navigation commands to the rover.

The computer and other equipment on board gets hot but the rover itself can get very cold. The MSL engineering task was to design a vehicle that could be used in any region of Mars, places in the polar regions where temperatures plunge to –209°F (–133°C), or equatorial regions where temperatures soar to 81°F (27°C). NASA wanted Curiosity to be a new-generation vehicle that could serve as a template for future rovers going to different places on Mars. And because the specific selection of a landing site for Curiosity would come long after development of MSL had been completed, it wanted to give scientists the widest possible range of potential sites to choose from. The thermal challenge facing MSL design engineers also included, for the first time on a Mars roving vehicle, heat generated by the radioisotope generator (RTG) installed for electrical power production.

The WEB is built to maintain internal temperatures between –40°F (–40°C) and 122°F (50°C). Legacy technology from earlier vehicles was only so good at tackling the problem of maintain a suitable thermal balance because of the added demands of MSL. But the principle remained the same, Curiosity has a pumped fluid loop running though an avionics mounting plate with a thermostatically controlled balance between the heat generated by the RTG and passively vented to the atmosphere of Mars by its cooling fins, and the heat needed to warm the WEB. Small heaters are provided for warming areas where there is insufficient heat to keep components sufficiently warm. Above the main body of the rover, the Remote Sensing Mast has such a heater.

Electrical power for Curiosity is the Multi-Mission Radioisotope Thermoelectric Generator (MMRTG), a development of numerous RTGs launched over the last several decades for providing power to space systems. They work on the principle of conversion of heat into electricity, discovered in 1821 by the German scientist Thomas Johann Seebeck. In this, two dissimilar electrically conductive materials are joined in a closed circuit and the two junctions are kept at different temperatures. These junctions, in pairs, are called thermoelectric couples (thermocouples), where the output in energy is a function of the different temperatures

and the properties of the thermoelectric materials. The cold junction is the cold of the vacuum in space while the heat source for the hot junction of the thermocouple in an RTG is Plutonium-238, which has a melting point of 1,742°F (950°C), a half-life of 85 years and emits primarily α-particles. The proportionality constant is known as the Seebeck coefficient.

RTG power sources can provide stable, consistent electrical energy night or day for very long periods of time at less weight than equivalent sources. They had first been used on a Mars mission when the two Viking landers touched down in 1976. Curiosity's 99lb (45kg) MMRTG is about 25in (64cm) in diameter across the fin tips, has a length of 26in (66cm). It is assembled from eight General Purpose Heat Source (GPHS) modules, containing a total of 10.6lb (4.8kg) of Pu-238 dioxide providing an initial 2kW of thermal power and 120W of electrical energy. Balancing the thermal requirements of the different parts of the rover is handled by the Mechanically Pumped Fluid Loop (MPFL), which is based on a heat rejection and recovery system comprising two Freon (CFC-11) loops that interact with each other to provide total thermal management.

Known as the Rover HRS (RHRS), the first loop consists of a set of pumps, thermal control valves and heat exchangers (HXs) responsible for transporting heat from the MMRTG to rover electronics on cold days or straight to the environment for warmer periods. The second loop, known as the Cruise HRS (CHRS), would have thermally balanced the system during the long journey to Mars and could more directly dissipate heat from both the rover avionics and the cruise stage directly to the cruise stage radiators. A unique system was developed for Curiosity enabling both heat collection from the MMRTG and rejecting waste heat to the environment, consisting of two honeycomb core sandwich panels with HRS tubes bonded to both sides.

Two such HRS units surround the MMRTG at the aft end of the rover so that heat acquired on the inner surface is rejected on the outer surface. To operate efficiently the fluid tubes in the two surfaces need to be isolated from each other, and the HXs are designed for high in-plane thermal conductivity and extremely low through-thickness conductivity by installing aluminium facing sheets and aerogel inside a honeycomb core. Very precise and complex assemblies of bent aluminium tubes are carefully bonded by hand on each side of the HX panels.

The MMRTG was designed and provided by the US Department of Energy to a new specification that demanded a 14-year life, but the great advantage for Curiosity is that unlike

ABOVE Multiple images taken by the mast camera of Curiosity provide detail on the top decking with circular sample collection shafts to the right, the crossbeam supporting the mobility system and the support structure and radiator panels for the MMRTG electrical power source at the rear. *(JPL)*

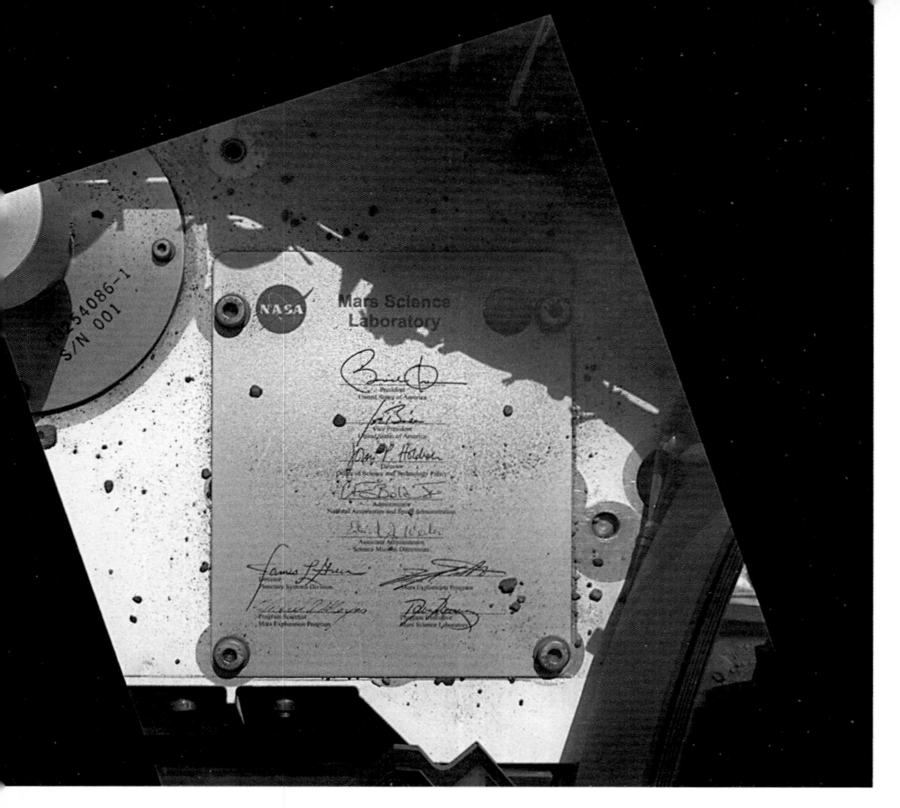

ABOVE A plaque placed on the front left of the top deck of Curiosity bears the signature of the President and the Vice-President, John Holdren the director of the Office of Science and Technology Policy, Charles Bolden the NASA administrator and Ed Weiler, associate administrator for NASA's Science Mission Directorate. Taken on the surface of Mars, the plaque and top decking contain deposits blown up into the air during the landing phase. *(JPL)*

solar cells used on the previous rovers, this electrical power source is not degraded by lack of Sun, hazy skies or dust settling on solar-cell arrays. Electrical output from the MMRTG charges two lithium-ion rechargeable batteries, each of which has a capacity of 42 amp-hours and will experience several charge–discharge cycles each Martian day. Power requirements for rover systems and equipment places great demands on the electrical production and distribution system and the physical mobility of the six-wheel drive system as well as the RSM and the robotic arm add to those demands.

The Robotic Arm (RA) is an articulated arm attached to the front of the rover body, extending from that point a length of 6.2ft (1.9m) at full reach. It supports a turret 23.5in 60cm) across, supporting a variety of tools and equipment for conducting surface science. Designed and manufactured by the Space Division of MDA Information Systems, specialists in building robotic arms and manipulators for the Shuttle and the International Space Station, it was fabricated in Pasadena, California, near JPL. The RA has five degrees of freedom, with rotary actuators at the shoulder azimuth joint which attaches it to the rover, a shoulder elevation joint, elbow joint, wrist joint and turret joint. Between the joints are structural elements with long links connecting the shoulder and elbow joints, the upper arm link, and the elbow and wrist joints known as the forearm link.

The turret is essentially the 'hand' to the arm and carries, as fixed instruments, the Mars Hand Lens Imager (MAHLI), the Alpha Particle X-ray Spectrometer (APXS), components for the Sample Acquisition/Sample Processing and Handling (SA/SPaH) subsystem, the Powder Acquisition Drill System (PADS), the Dust Removal Tool (DRT) and the Collection

RIGHT The heat shield about to be installed. Curiosity will not get its nuclear power source until it is on the launch pad, the housing here seen vacant. *(JPL)*

and Handling for In-Situ Martian Rock Analysis (CHIMRA). The working area and volume for the RA is best described as an imaginary cylinder, known as the robotic arm workspace. It is 31.5in (80cm) in diameter, 39.4in (100cm) high and positioned 41.3in (105cm) in front of the rover body, extending 7.9in (20cm) below the surface.

The MAHLI and APXS instruments are contact devices while the PADS, DRT and CHIMRA are for sample preparation and preparation functions, all described in the 'Science on Mars' section below. The contact equipment on the RA turret is required to exert considerable force against a rock or boulder, and in order to stabilise the instrument and prevent it from 'walking' across the surface the RA actuators allow the mechanism to apply a force of 240–300N in certain arm configurations. In motion the tip speed of the RA is about 0.4in/sec (1cm/sec).

ABOVE Under armed guards discreetly out of sight, the highly toxic Plutonium-238 MMRTG is hoisted to the top of the launch pad for installation. *(KSC)*

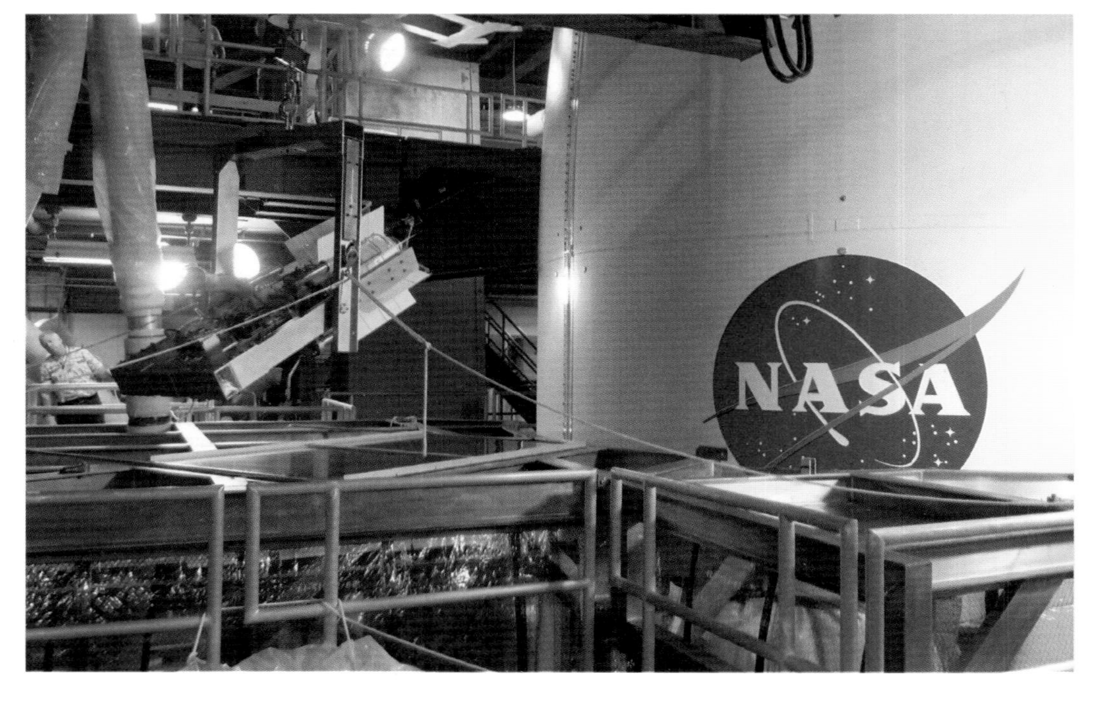

LEFT Hoisted by a special lifting device, the nuclear power source complete with cooling fins on the MMRTG is manoeuvred into place through the side of the payload fairing, through the aperture in the aeroshell and on to its fixture in the back of Curiosity. *(JPL)*

Curiosity has high demand for electrical power and for communications with Earth, either direct via the High-Gain Antenna (HGA) or relayed through orbiters around the planet. Two antennas operate in the X-band range (7–8GHz) while a third operates as an orbiter-relayed link via a UHF antenna (400MHz). The X-band system uses a 15W, solid state power amplifier to feed the HGA, hexagonally shaped and 11in (28cm) across, attached to the left rear section of the top deck on the main body of the rover.

The HGA with its 48-element microstrip path radiating element of 10 x 11.6in (25.5 x 29.4cm) was provided by EADS, the European Aeroneutronic Defence and Space company. At 5° off-boresight it has a downlink gain of 4dB lower with an uplink gain 3dB lower than nominal. The X-band system operates at a downlink frequency of 8,395MHz and an uplink frequency of 7,145MHz. It can transmit data at a rate of at least 160 bits/sec to the 112ft (34m) diameter DSN antennas on Earth or at a rate of at least 800 bits/sec to the 230ft (70m) dish antennas.

In normal operations a typical daily uplink takes 15 minutes for a total volume of 225kb of data at a nominal rate of 1–2 Kbits/sec, and this provides time for packaging the commands into groups and for transmitter delays from Earth. The X-band low-gain antenna (LGA) is omnidirectional with a broad footprint, and therefore does not require pointing at the Earth. It is situated behind the HGA on the same side of the rover. The LGA can uplink communications at 15 bits/sec at maximum range between Earth and Mars.

BELOW **Curiosity has a geometry that permits all-round vision for the six mast-mounted cameras, and the articulation of the masthead is designed to afford full panoramic vision.** *(JPL)*

The single-quad UHF antenna is located on the opposite rear side of the rover on a special mounting bracket and is used with either one of two redundant Electra-Lite radios, a smaller version of the Electra radio flown aboard Mars Reconnaissance Orbiter. Cleverly, when used as a relay the MRO radio commands Curiosity's UHF radios to switch data rates according to the strength of the received signal. The MRO–MSL relay link can also be used in 'safe' mode to communicate a wake-up call to the rover, should it, for instance, 'lose' its clock timing and be uncertain as to when to call up MRO for a data link. But the UHF link can also be used through the Odyssey spacecraft.

Three standard UHF frequency channels are available for the relay link: channel 0 (437.1MHz) exclusively for links through Odyssey; and channel 1 (435.6MHz) and channel 2 (439.2MHz), either of which would normally be used for communicating through MRO. Seven selectable uplink rates are available, at 2, 8, 16, 32, 64, 128 and 256Kbps. Only 8 and 32Kbps rates are used by Odyssey.

Eyes on Mars

Curiosity carries a wide range of camera equipment, providing scientists and engineers on Earth with eyes on Mars. Apart from the scientific duties assigned to the rover on a continuous basis through programmed computer commands, there was also a need for autonomous operation that demanded copious information for navigation and warning of hazards that could imperil the vehicle. Accordingly, cameras were assigned either 'science' or 'engineering' roles, although in operation there would be an overlap between the two functions. Curiosity is equipped with 3 science cameras and 12 engineering cameras. Two science cameras known as MastCam play a peripheral role in securing images of near, mid- and far field views which would also be pivotal in planning traverses across the surface as well as giving geologists a direct view of the rover's surroundings. A third science camera is a small optical device for examining rocks, surface soils and particles of dust close up. The 12 engineering cameras were needed for navigating around and for providing warning of potential hazards in the path of the vehicle.

Provided by Malin Space Science System, the two MastCam cameras mounted to the bar on the Remote Sensing Mast (RSM) beneath the ChemCam instrument are approximately 2m above the surface of Mars and each is capable of acquiring images in natural colour and visible/near infrared multispectral views, the latter a vital tool for deciphering the mineralogy of surface samples and features. These cameras can also be used to construct high-definition stereo views of the surface surrounding Curiosity. The two cameras each come with a camera head and a Digital Electronics Assembly (DEA), the latter situated alongside DEAs for MARDI and MAHLI in the thermal comfort inside Curiosity. Both cameras have a mechanical and an autofocus capability and can focus from 2.1m to infinity.

Known as 'MastCam 100', one camera has a 100mm focal length f/10 lens providing images with a scale of 7.4cm/pixel at a distance of 1km and about 150 microns/pixel at 2m. The square field of view covers 5.1° over 1,200 x 1,200 pixels on the 1,600 x 1,200 CCD. The other camera, dubbed 'MastCam 34', has a 34mm focal length f/8 lens and the 15° fov covers the same pixel area and CCD area. This camera can capture 450 microns/pixel at 2m and 22cm/pixel for subjects at a distance of 1km. Both cameras use a Bayer Pattern Filter to avoid the need for taking three separate images with three filters (red, green and blue, or RGB), saving on time and data storage. Uniquely, the camera used on Mars Global

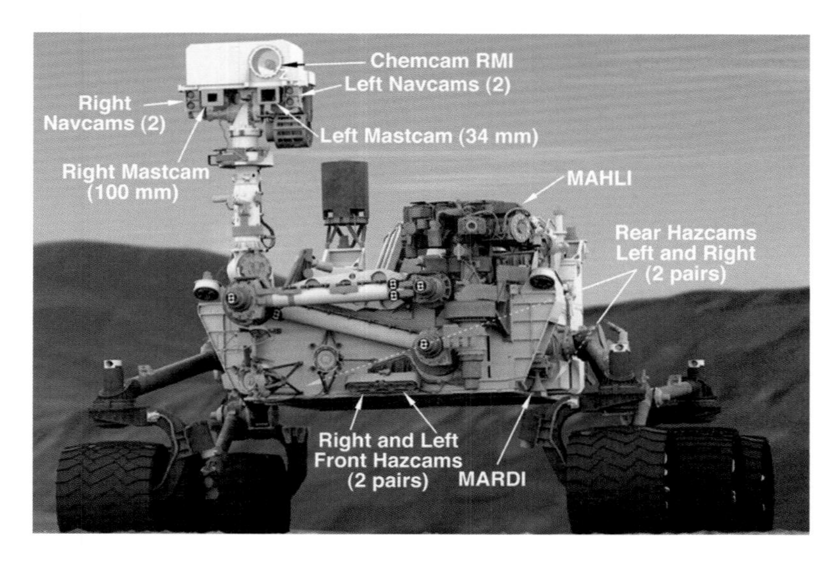

ABOVE Curiosity's cameras are applicable to both science and engineering tasks, their individual and unique specifications aiding in both scientific research and for navigation and hazard warning to the autonomous control systems. *(JPL)*

LEFT Provided by Malin Space Science Systems, the 100mm MastCam 100 affords good long-range views of the surface. *(MSSS)*

RIGHT **The Malin MastCam 100 with its 100mm telephoto lens.** *(MSSS)*

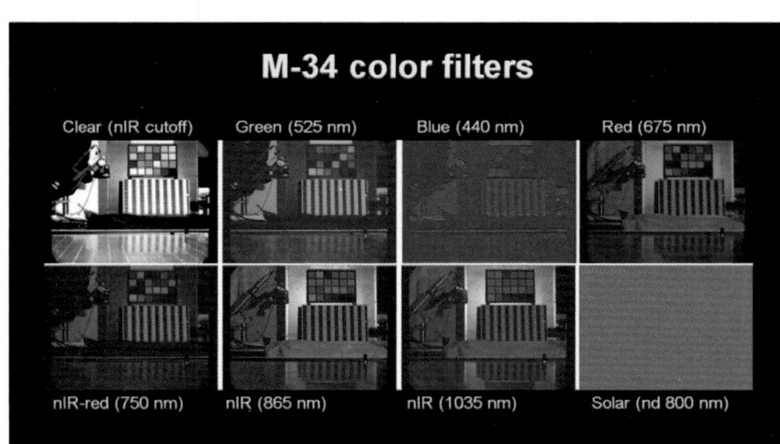

BELOW **MastCam 34 camera images obtained through the eight optional filters. Image at upper left shows a Bayer colour array very close to that obtained from a consumer digital camera. Other images show narrow exposures of about 20–25nm through different filters to build a spectral view. The apparently blank image is obtained through the neutral density filter for direct observations of the Sun.** *(JPL)*

Surveyor used red and blue filters, the green being an interpolation based on the known relationship between these two to obtain green.

Each MastCam camera has a filter wheel covering different, individually selectable narrow visible and near infrared wavelengths. In nanometres, for the 34mm lens these are 440, 525, 550, 675, 750, 865, 1,034 and 440 neutral density. For the 100mm lens these are 440, 525, 550, 800, 905, 935, 1,035 and 880 neutral density. The two neutral filters would be used for views of the Sun and each wheel has an additional IR-rejection filter for RGB imaging using the Bayer Pattern CCD. Filter changes to consecutive positions take 5–8 seconds and a full wheel rotation takes 30–45 seconds. Each camera also has the capacity to obtain 720p HD video (1,280 x 720 pixels) at 10 frames/sec.

It takes less than one hour to obtain a full panorama (360° x 80°), and with an internal storage capacity of 8GB the images can be stored indefinitely or erased on command. This buffer space allows storage of up to 5,500 RAW images and JPEG compression is available while the footage is being taken. The usual process is to transmit thumbnails of all images, sent to Earth at 200 x 150 pixels so that scientists can select those of sufficient interest to warrant the full high-resolution versions to be sent down. The system can also acquire stereo-pairs for 3D imaging but this is not a high priority due to the different focal lengths of the two lenses. But compared to the two MER rovers, MastCam is a major step forward. The MER Pancam acquires 1,024 x 1,024 pixel images covering 16° x 16°. This means the 34mm MSL MastCam images have about 25% greater spatial resolution while the 100mm camera has 3.67 times the spatial resolution of Pancam.

The technical capability of the MastCam cameras is a product of the requirements issued by the principal investigator and was essential for the work Curiosity was required to do. When MSSS received the contract to build MastCam, it was designed to comprise two identical area-array digital cameras with 15:1

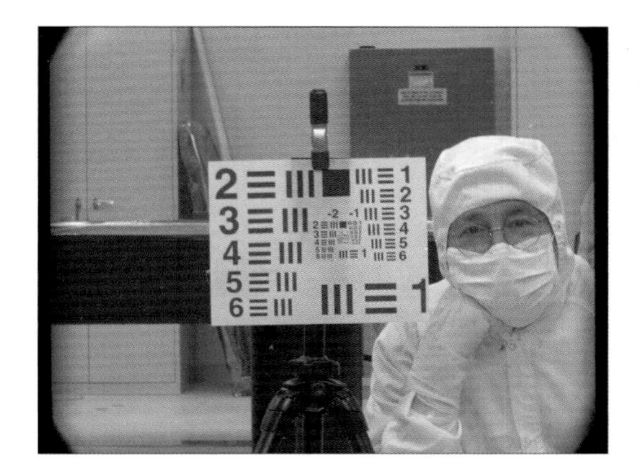

ABOVE Michael Malin of Malin Space Science Systems poses for a test of MastCam 34 with a 1951 US Air Force Resolution Chart, a test target which here is demonstrating that the camera has a resolution of 4.5 lines/mm at a range of 6.2ft. *(MSSS)*

zoom telephoto lenses. The electronics were to be identical to those on MARDI and MAHLI cameras, also provided by Malin. They would have had the same focal length binocular vision with 14 filter wheel positions for multispectral analysis of surface features.

But when technical and budget problems hit the MSL programme and threatened to derail the project, in 2007 a senior executive at NASA Headquarters ordered a reduction in the capability of the MastCam system by eliminating the zoom and focus capability, despite there being no indication that it was one of the instruments running into financial problems; by cutting back the budget on MastCam he hoped to absorb cost inflation from other parts of the programme. Robust lobbying reinstated the focus capability, without which the MastCam would have been far less productive for the science teams.

The principal science investigator for the Mars Exploration Rover, Steve Squyres, insisted on Spirit and Opportunity having a Microscopic Imager (MI) on each spacecraft, just like any geologist on Earth would carry. His successor on MSL was determined to have a similar capability, so Malin Space Science Systems provided the Mars Hand Lens Imager (MAHLI) for Curiosity. Mounted on the turret at the end of the robotic arm, it would provide close-up images of samples to document them in-situ and help define the nature of surface materials and fine-scale structures. The outstanding advantage afforded to geologists by the MI on the two MER rovers was repeated on Phoenix and would be advanced to another level by the MAHLI on Curiosity.

Like other cameras on MSL, the camera head is remote from the Digital Electronics box, situated down inside the rover's WEB. And, once again, the camera uses a Bayer Pattern Filter CCD for a balance between natural colours. The specific CCD fitted to MAHLI provides a 1,600 x 1,200 pixel-sized image and so is essentially a two-megapixel camera. It incorporates a motor for adjustment of lens positioning and for focusing on a specific target, operating at distances between 22.5mm and infinity. Close-up images have a resolution of 15 micrometres/pixel covering an area of 18 x 24mm. At a working distance of 50mm each image has a resolution of 24.5 micrometres/pixel. At 66mm the 31 micrometre/pixel resolution is about the same as that from the MI on the MER rovers but covers a larger area and provides images in colour. The camera can also be used to build a dimensional map of the area imaged through focal plane merging (or z-stacking), where several separate images are taken at different focus positions.

ABOVE The Mars Hand Lens Imager (MAHLI) provides close-up views of rocks and samples and has a resolution up to 14 micrometres/pixel with a focal length adjustable for imaging from a distance of 0.8in to more distant objects. *(JPL)*

RIGHT The MAHLI calibration target incorporates a metric standard bar graphic, a US penny, a stair-step pattern for depth calibration and is 8.5in by 11in in size. Patches of pigmented silicone include five left over from the MER programme. The penny is from 1909, the centennial of President Lincoln's birth. Numbers on the standard US Air Force chart indicate the number of black-and-white bars that will fit into a 1mm space. *(JPL)*

The MAHLI camera also has four white-light LEDs and two ultraviolet LEDs for night-time illumination or for objects in hard shadow. By using the robotic arm to move between two images, stereo or 3D images can be taken, thus enhancing the ability of the scientists on Earth to use their eyes as they would on the surface of Mars. Because of the close proximity of the lens to the surface of targets, and because Mars can be a very dusty place, a fine-particle screen covers the optics but has a window through which pictures can still be taken, should conditions prohibit exposure of the lens.

The engineering cameras on Curiosity comprise four navigation cameras, known as Navcams, and eight hazard warning cameras, or Hazcams. They were built at JPL by the same people who built similar cameras on the two MER rovers but incorporated more powerful heaters to lower the temperatures at which they could operate on the surface. Like the science cameras, each comprised a detector/optics head and an electronics box inside Curiosity. Technicians at JPL built a total of 26 cameras, including the 12 flight units, 4 flight spares and 10 engineering units used for test and evaluation. The electronic boxes are each 2.64 x 2.72 x 1.3in (67 x 69 x 34mm) in size.

The four Navcams are set high up on the Remote Sensing Mast at the approximate height of the MastCams to give a long view, while the eight Hazcams are low down on the rover body. The Hazcams were the first to send back images from the surface and were used to plan the first excursions. At the end of each drive images were taken and transmitted to Earth for precision updates on the rover's exact stopping point and on the view from that new location of navigable routes through obstacle or potential hazards to plan for in the next set

RIGHT The MAHLI camera located at the turret end of the robot arm refers to the calibration target mounted on the shoulder joint. *(JPL)*

of traverse commands. Engineering images are also used to position the robotic arm for the delivery of material to the rover's onboard sample processing system.

Derived from the equivalent MER cameras, the Navcam cameras use f/12, 14.67mm fixed focal length lenses which provide a 45° x 45° field of view (60.7° on the diagonal) and a scale of 0.82 millirad/pixel at the centre of the fov. Depth of field ranges from 0.5m to infinity, with a hyperfocal distance of 1m. The Hazcam cameras have f/15, 5.58mm, fixed focal length f-theta fisheye lenses providing a 124° x 124° field of view (180° on the diagonal). The scale at the centre of the image is 2.1 millirad/pixel, with a depth of field of 4in (10cm) to infinity and a hyperfocal distance of 20in (50cm).

Both types of engineering camera have bandpass filters of 580–800 nanometres. Stereo baselines are 16.7in (42.4cm) for the Navcams, 6.6in (16.7cm) for the front Hazcam and 3.9in (10cm) for the rear Hazcam. Boresight pointing for the Navcam is 0–360° in azimuth and –87° to +91° in elevation. Both front and rear Hazcams have pointing set at 45° below horizontal. From their position atop the RSM, the two Navcams are situated 75in (1.9m) above the ground, the front four Hazcams at a height of 27in (68cm) and the rear four Hazcams at 31in (78cm). With a detector head only 41 x 51 x 15mm in size, each Navcam weighs 0.45lb (220g), each Hazcam 0.55lb (245g).

The engineering camera detectors were put together and packaged up using spare MER CCD wafers, the detector being a frame-transfer device with an imaging region of 12.3 x 12.3mm of 1,024 x 1,024 pixels, each pixel being 12^2 microns. Readout time is 5.4 seconds for a full frame or 1.4 seconds for a binned frame and a full readout time less than 5.4 seconds.

The Remote Sensing Mast supports the two Navcams, the two Pancams, the ChemCam science instrument (see 'Science on Mars' section) and the Rover Environmental Monitoring Station (REMS). It incorporates a heated compartment for the cameras and the ChemCam instrument known as the Remote Warm Electronic Box and the top of the RSM is about 1m above the top deck of the rover. The upper section is pivoted in two axes and has commanded motion through a full azimuth (362°) and a sweep of 178° (+91°/–78°) in elevation, with an additional 4° in elevation allowed to prevent stress on the cables.

The RSM can position the four cameras carried on the masthead to take images of the top of the rover deck, the mobility system, the robotic arm workspace and the sky. Absolute pointing accuracy is 46 milliradians, equivalent to 6 Navcam pixels. Motion commands are sent for respective azimuth/elevation steps or in three-dimensional Cartesian target points in a specific coordinate reference frame. With the four Navcam cameras 6.2ft (1.9m) above the Martian surface, they have a flat unobstructed view to the horizon 12 miles (19km) away.

Updates within the navigational frame of reference for plotting traverse routes demands

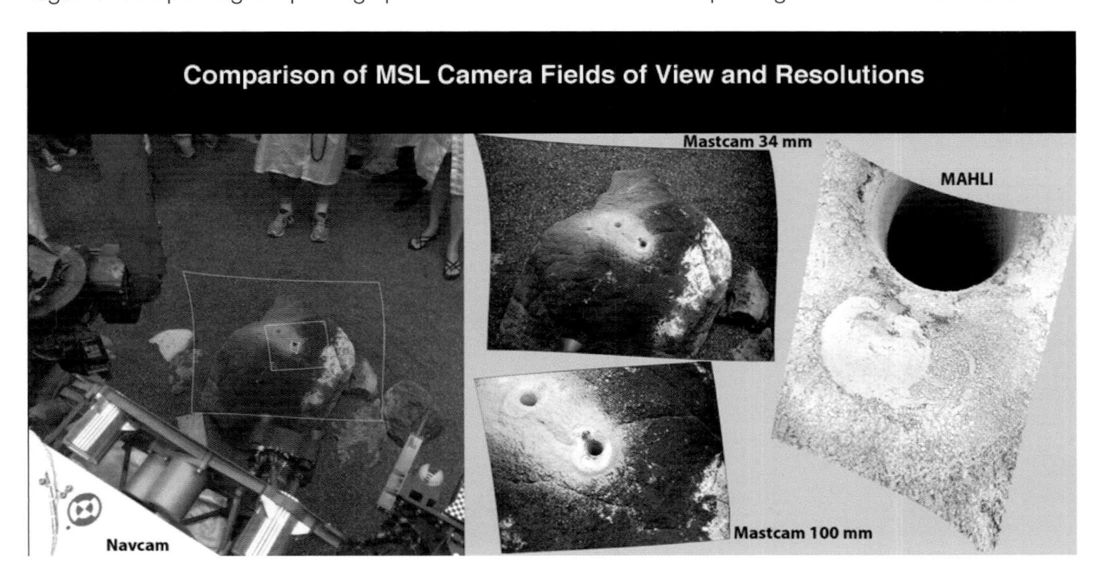

LEFT Test images from four cameras. From left: a black-and-white image from a Navcam with a field of view of 45°; MastCam images at 15° and 5°; a MAHLI image at 23°. The rock target is on the right of the picture. *(JPL)*

RIGHT The Digital Electronics Assembly that provides an interface with the MastCam, MAHLI and Mardi cameras on Curiosity. *(MSSS)*

a highly accurate knowledge of the direction in which Curiosity is pointing, and in the absence of an Earth-like magnetic field this requires careful plotting through known angles of azimuth and elevation from the RSM/camera system. An onboard software ephemeris system known as Inertial Vector Propagation (IVP) has been developed from the Pathfinder and MER era and provides Curiosity with its own highly accurate plotting frame.

Through the IVP, the RSM is driven around until the two Navcams are pointing directly at the Sun. Although not designed to take images of the Sun – even an exposure time of 5.12 milliseconds would overexpose the image – they can be safely used in this way to obtain pointing knowledge. From this azimuthal reading flight controllers can update the angular pointing of the rover as it sits on the surface.

The imaging flight software for the 12 Navcam and Hazcam cameras runs through the Rover Compute Element, Curiosity's main computer, divided into two separate functional components known as A and B, one being backup to the other on a selectable basis. Each RCE is connected to a dedicated set of six engineering cameras, and the two are not cross-strapped. Only a single RCE can communicate with its string of cameras, and to obtain images from the other set of cameras the other RCE must be enabled. In operational use, generally one RCE will be shut down while the other acts as the prime. Switching between the two is both unusual and unwise.

Imaging flight software is another evolutionary development from the Pathfinder/MER spacecraft, and software command parameters include: manual and autoexposure, exposure timetable storage, exposure time scaling, histogram generation, row/column summation, thumbnail generation, 12- to 8-bit compounding, spatial down-sampling, spatial subframing, shutter subtraction, bad pixel correction, flat field vignette correction, geometric camera model management, stereo processing and image metadata collection.

RIGHT A full frontal view of the Hazcams and the four optical cameras on the RSM. *(JPL)*

The software uses the ICER wavelet image compressor for lossy image compression and the LOCO image compressor for lossless compression of the image. Both Navcam and Hazcam images are obtained through a specified series of commands for image acquisition and processing while Navcam panoramas are ordered through a series of individual image commands. The flight software commands camera power on and off with up to two cameras powered simultaneously. Images are buffered in a non-volatile memory/camera interface card and from there transferred to the appropriate RCE for processing, storage and downlink.

Engineering cameras are given unique designations according to their location. The four Navcams on top of the RSM are designated to left or right of the rover looking forward in the direction of travel. Situated in pairs one above the other either side of the two Pancams, the upper pair are wired to the RCE-A computer and the lower pair are wired to the RCE-B element. Each Navcam pair has a stereo baseline of 17in (42.4cm) while the two 'A' cameras are mounted 1.9in (4.8cm) above the 'B' cameras. These are able to observe all three wheels on the right side of the rover, but due to physical obstructions only the middle and rear wheels and a partial view of the front wheel on the left side. There are no lens covers on the Navcams, instead they are stowed in a protective hood until deployment after landing. After the RSM is deployed they can be pointed downward to prevent dust settling on the lenses.

The eight Hazcams are immovable, fixed in their mounting to the front and rear endplates of the rover body. The front four cameras are situated in horizontal pairs and are designated as 'right' and 'left' for each side of the rover. Each pair on the right (as viewed from a hypothetical driving position on the rover looking forward) is assigned respective elements of the RCE computer, the outboard Hazcam on the right side being wired to RCE-A, the adjacent inboard Hazcam to RCE-B. On the left side of the front of the rover, the outer Hazcam is wired to RCE-B, the inboard one to RCE-A. They are interleaved horizontally with a 3.2in (8.2cm) offset with a 6.5in (16.6cm) stereo baseline. All eight cameras are 26.8in (68cm) above the surface.

The rear Hazcam pairs sit each side of the Radioisotope Thermoelectric Generator, and are farther apart than the front pairs, with a stereo baseline of 39in (1m). They sit 31in (78cm) above the surface but a similar designation of each camera is applied as that used for the front cameras. The two cameras on the left side of the rover (looking at the back and facing

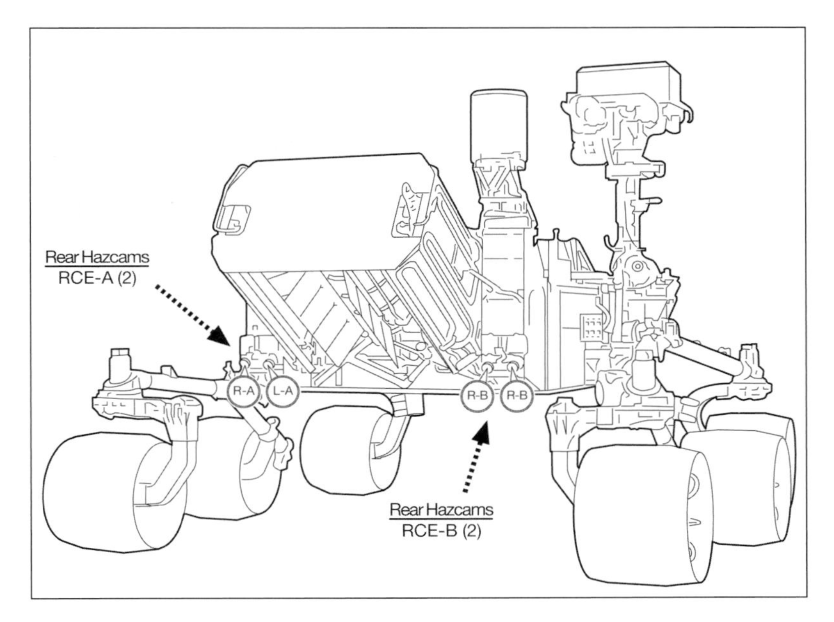

ABOVE The rear hazard cameras on Curiosity, divided for command and data between the two RCE elements of the computer. *(David Baker)*

BELOW Forward Hazcams wired to respective computer segments together with the RED affiliations for the Navcams on the Remote Sensor Mast. *(David Baker)*

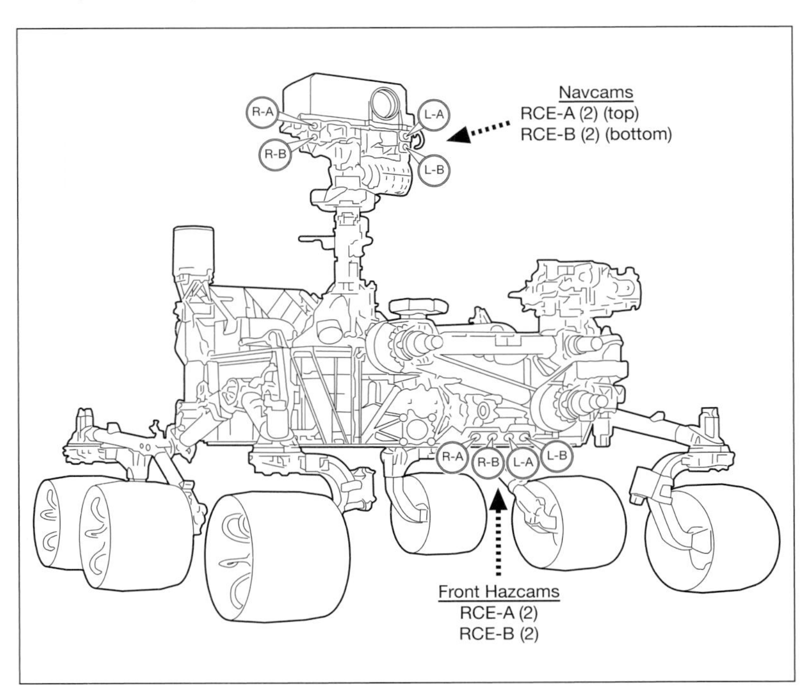

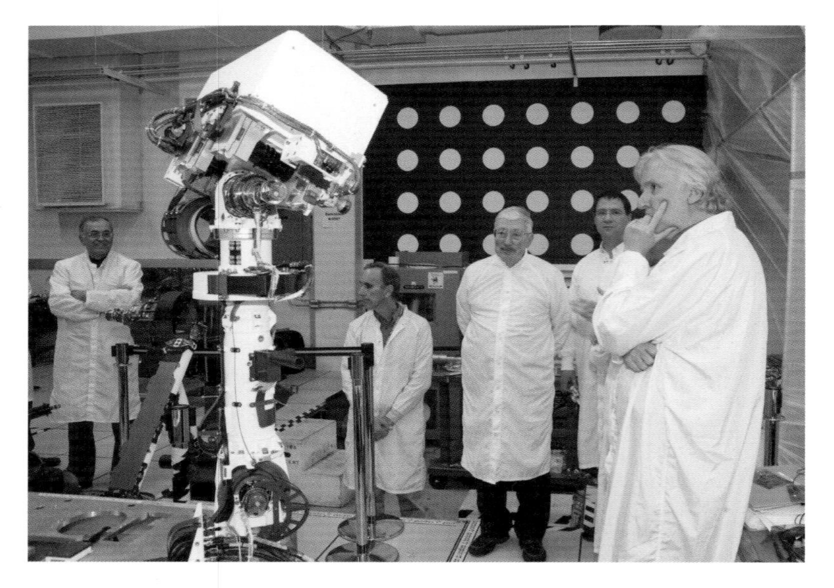

ABOVE James Cameron (right) meets key players in the Mars Science Laboratory mission including (left to right) JPL director Charles Elachi, MSL project scientist John Grotzinger, MSL programme manager Pete Theisinger, and Michael Malin of Malin Space Science Systems. *(JPL)*

forward) are designated 'right' (outer) and 'left' (inner) and both are wired to RCE-A. The other pair, similarly designated, are routed to RCE-B. The 'A' cameras can view the left rear wheel but not the right, and vice versa for the right 'B' pair. Of course, it is possible to see both wheels by switching computer elements. All eight Hazcams have transparent covers that are opened after initial checkouts on the surface.

Because the stereo baselines are greater than those for the engineering cameras on the MER rovers, the MSL Navcams and front Hazcams provide stereo images that are 2.1 and 1.7 times more accurate than those transmitted by Spirit and Opportunity. The rear Hazcams have the same stereo separation as those on the MER rovers. During operations, a complete 360° panorama consists of 12 stereo pairs spaced at 30° in azimuth, a spacing less than the 45° Navcam field of view but which allows a more efficient overlap. The MER Navcams used a 36° spacing for panoramas but the 17in (42.4cm) stereo baseline on MSL versus 7.9in (20cm) on the MER rovers allows better inter-image spacing.

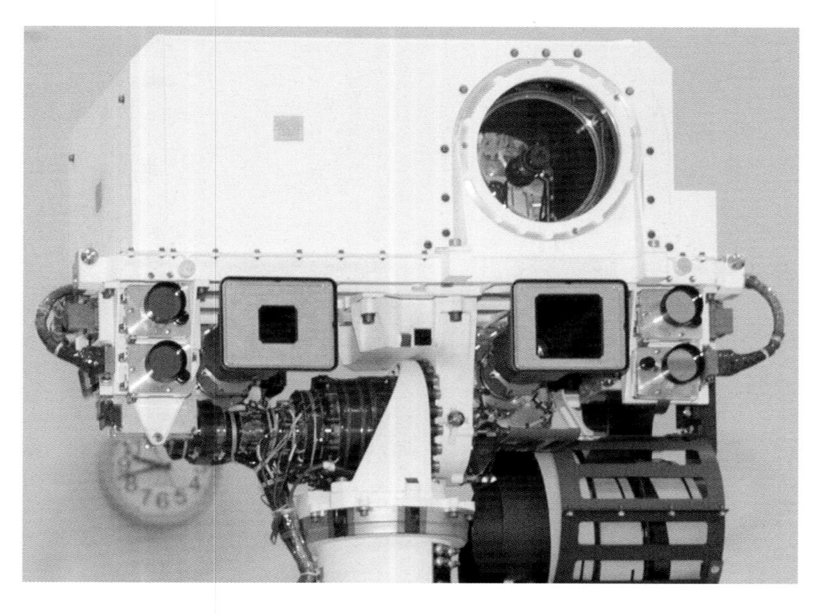

LEFT The ChemCam instrument atop the head of the RSM consists of a laser and optical telescope that will conduct remote sensing of rocks by firing a beam at a target on the surface and using a spectrometer to record the elemental composition. *(JPL)*

Science on Mars

Curiosity carries ten science experiments, each of which has several elements or components designed for specific functions but all seeking to answer the fundamental question that gave rise to the project: were there ever, or are there now, environmental conditions conducive to the origin and development of life. Curiosity is not in itself a search for life and that will probably never be the function of a single mission. With a total mass of 165lb (75kg), the science payload carried by Curiosity is almost seven times the weight of the five instruments carried by each MER rover.

The cameras reviewed in the previous section are but one part of the cluster of science instruments, although as described they share functionality with engineering requirements. Now we need to look at the rest of the equipment, starting at the top of the Rover Sensor Mast, migrating down to the robotic arm and then to the interior of the rover itself. It is in this group that Curiosity, like the optics, excels.

Making its operational debut on a planetary mission of any kind, the Chemistry and Camera instrument, or ChemCam, was designed to conduct remote sensing of rocks and surface materials using a technique known as laser-induced breakdown spectroscopy. This technique has been used for examining materials in extreme environments, such as

LEFT ChemCam is used universally on Earth for determining the elemental composition of rocks but this is its first use on another world. The image shows the laser beam in visible light whereas in reality it is in the invisible infrared part of the spectrum. *(JPL)*

the sea floor or inside nuclear reactors, where it is almost impossible to put humans. It is integral with an instrument called the Remote Micro Imager (RMI), both of which are located on top of the Remote Sensor Mast above the compartment containing the MastCam and the two Navcams.

Known as the Laser-Induced Breakdown Spectrometer, the LIBS will provide scientists with the basic elemental composition of selected materials while the RMI concentrates its analysis on their geomorphological context. The ChemCam is able to conduct many more sample scans than can be reached with the contact instruments carried on the robotic arm and this helps direct scientists to rocks and places of particular interest, committing them to more detailed analysis by the other instruments. ChemCam is used on an almost daily basis to make elemental readings of soil and surface materials around Curiosity. It can also take readings of other places and rocks inaccessible to contact analysis with the arm turret.

The LIBS laser can hit rocks or surface targets up to 23ft (7m) away greater than 10mW/mm^2 exciting a 0.3–0.6mm spot ablating atoms and ions into light-emitting ionised gas, or plasma. Light from that spark is collected by the ChemCam telescope, which has a diameter of 4.33in (110mm), and focused into the end of a fibre optic cable. The 20ft (6m) cable carries the light down the RSM and into the main body of the rover where three dispersive spectrometers are located.

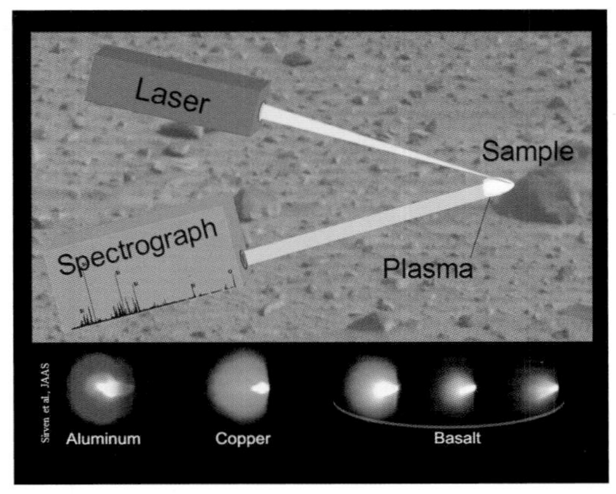

CENTRE A graphic illustration of how the Laser-Induced Breakdown Spectrometer is able to determine chemical content from spectral analysis. *(JPL)*

These record its intensity at 6,144 different channels (wavelengths of light) at 240–850 nanometres between the ultraviolet and the

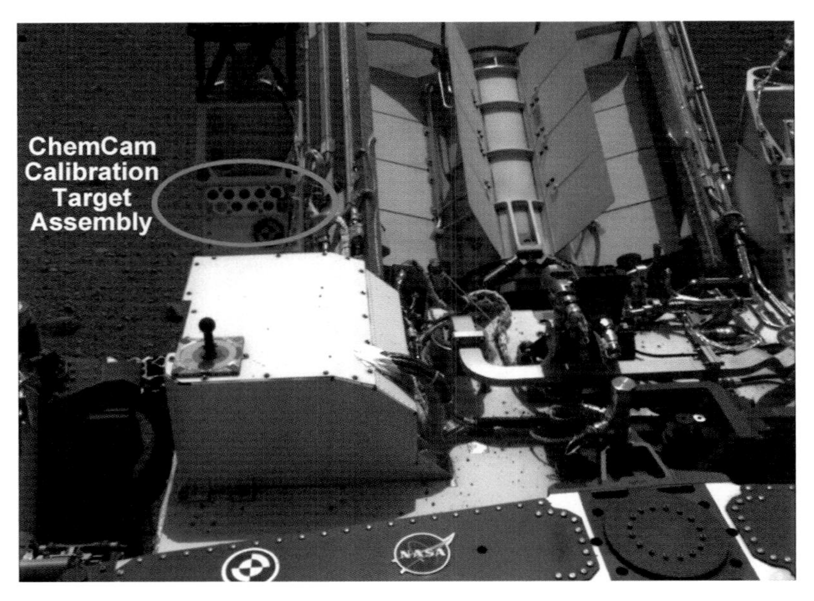

BELOW The ChemCam calibration target is located on top of the rover deck as seen in this view taken on the surface of Mars with the cooling fins of the nuclear (MMRTG) power source to the right rear. *(JPL)*

ABOVE A test of the ChemCam laser, where a pulsed laser beam is directed against an iron pyrite crystal in a sample chamber 10ft from the instrument. *(JPL)*

infrared. With a resolution from 0.09 to 0.30 nanometres in each channel the emission lines produce the data to determine the elemental composition, because different elements emit light at different wavelengths. About 50–75 blasts with the laser will, when the resulting spectra are overlaid, eliminate uncertainties to +/– 10%.

A wide range of elements can be determined by this method and scientists expect to be able to measure sodium, magnesium, aluminium, silicon, calcium, potassium, titanium, manganese iron, hydrogen, oxygen, beryllium, lithium, strontium, nitrogen and phosphorous. Several hundred repeated laser pulses can rid a sample of overlaying dust or grind away at a surface encrustation, allowing comparisons between the outer layer and the material beneath. ChemCam is also used to study the soil and, by using the RBI, measure the size and the distribution of particles over which the rover is travelling.

To provide the context in which the LIBS is recording spectra, the RMI can use the telescope to take monochrome images on the surface of a 1,024 x 1,024 pixel detector. Unlike the LIBS with which it operates, it has no distance limitation and has a field of view of 19

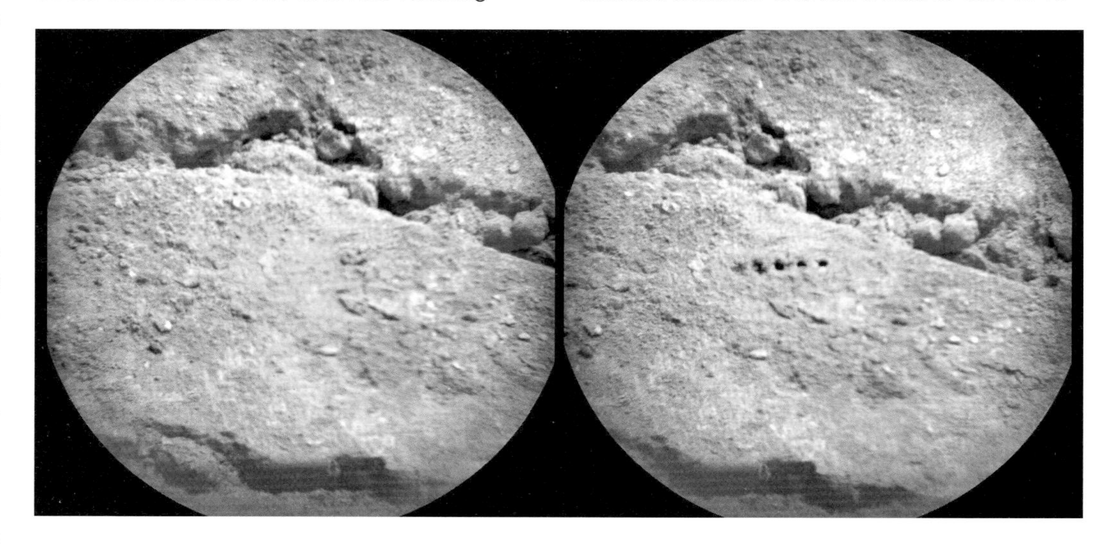

RIGHT Taken from a distance of 11.5ft on the surface of Mars during Sol 19, ChemCam's Remote Micro-Imager records the sputtering effect (right) of the laser with more than one million watts of power fired for five one-billionths of a second. *(JPL)*

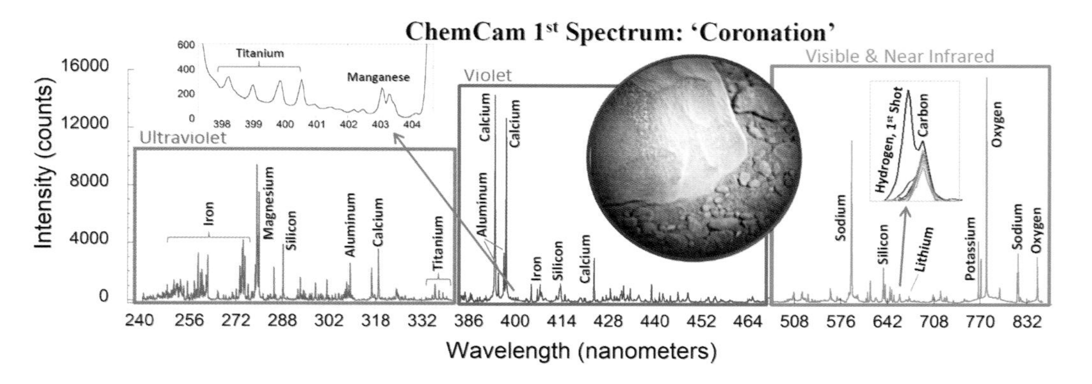

RIGHT The first spectral analysis made by Curiosity on the surface of Mars on 22 August 2012 following 30 laser shots at a single 0.016in spot on a rock indicating it to be basaltic composition. The inset at left shows the detail for titanium and manganese and a detail at right displays hydrogen and carbon peaks. *(JPL)*

LEFT The geometry of the robotic arm and its articulated arrangement of science equipment, sensors and cameras give Curiosity an unprecedented range of capabilities. *(JPL)*

milliradians, but, because the camera has been designed for LIBS, it has a resolution of about 100 microradians. From its vantage point 83in (2.1m) above the surface, co-aligned with the MastCam and Navcam, it can readily observe the submillimetre LIBS spot. Although the ablated area is microscopically tiny and difficult to detect, the RMI can find it at the operating distance of the laser. The RMI can be used independently of the LIBS for observing targets at any distance to infinity.

In preparing the ChemCam equipment, scientists were particularly excited about its ability to detect hydrated minerals indicative of bound water in minerals. It can also be used to measure the effects of weathering by successively sputtering away at the surface of a rock to analyse the migration of chemistry through the outer submillimetre layers. In several respects, the LIBS/RMI suite is a guide to structures of sufficient interest to dedicate several days of work using the contact instruments. The equipment has been a truly international endeavour. The French national space agency CNES funded ChemCam and the instrument was developed by the US Department of Energy at Los Alamos National Laboratory. A US scientist and a French scientist are principal and deputy principal investigators respectively. France built the ChemCam laser and the telescope itself.

The international flavour was maintained with an old Mars rover favourite, the APXS (Alpha Particle X-ray Spectrometer), a device provided by Germany for Spirit and Opportunity, and built for Curiosity by the Canadian Space Agency. Results from an APXS on Spirit and Opportunity provided crucial findings that went a long way to interpreting the sites at Gusev Crater and in the Meridiani Planum. APXS does the same job previously described for Sojourner and the two MER rovers, but for Curiosity there are some subtle improvements and an addition.

On Curiosity the sensor head is affixed as one of the instruments on the turret at the end of the robotic arm. Like its predecessors, it can determine the abundance of elements in rocks, including important rock-forming elements such as sodium, magnesium, aluminium, silicon, calcium, iron and sulphur. Because of the special nature of Curiosity's task, it is very sensitive to elements such as sulphur, chlorine and bromine that form salt. This can be a strong indicator of the presence of water at some point in the past.

Using Curium-244 sources for X-ray

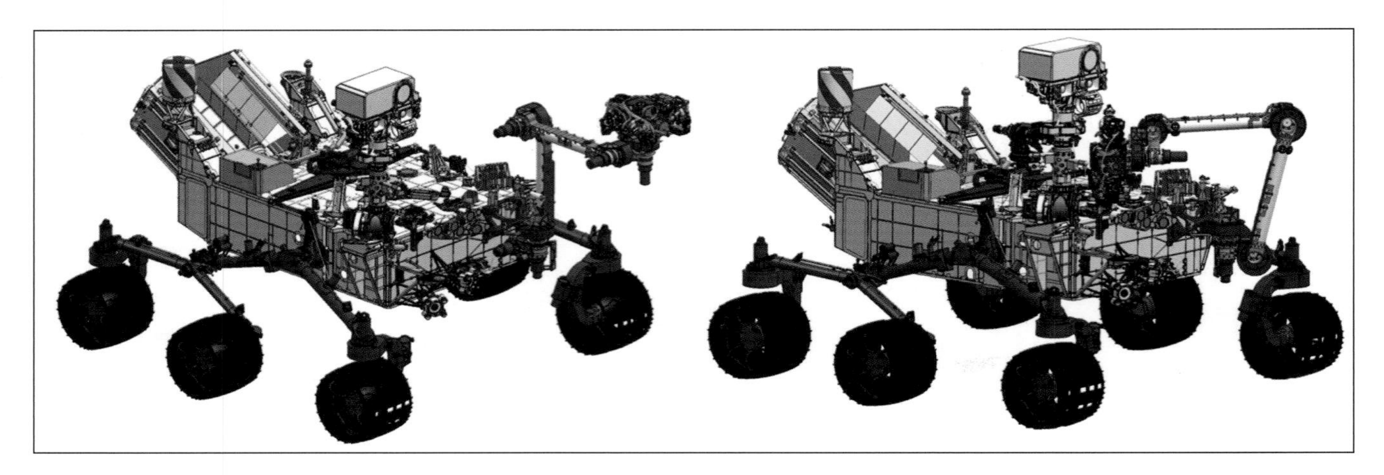

ABOVE The robotic arm has been designed to place the turret and its instrument suite anywhere on the surface up to 7ft in front of Curiosity. *(JPL)*

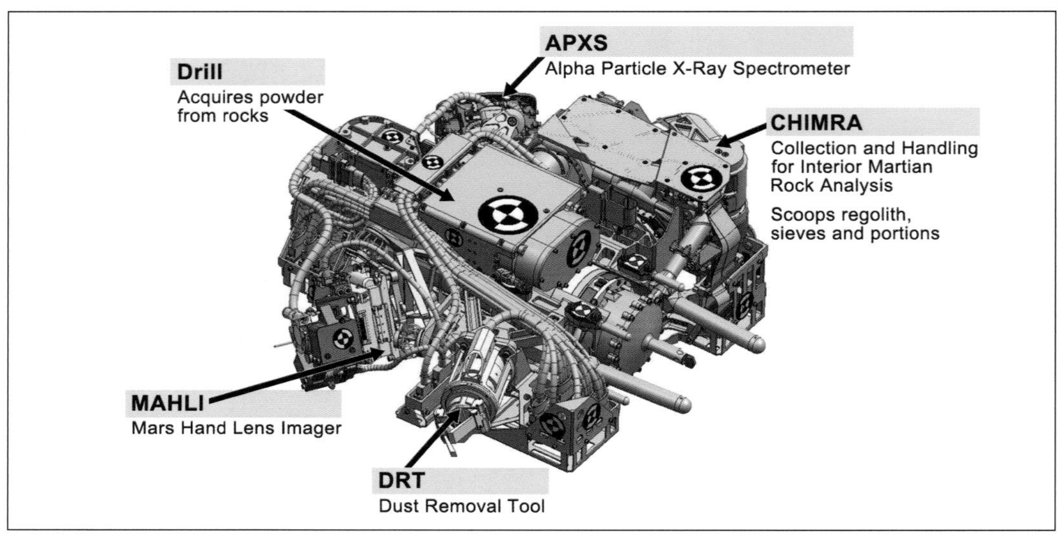

RIGHT The four primary instruments of the turret are all designed for contact science and for delivering samples to laboratories in the interior of the rover. *(JPL)*

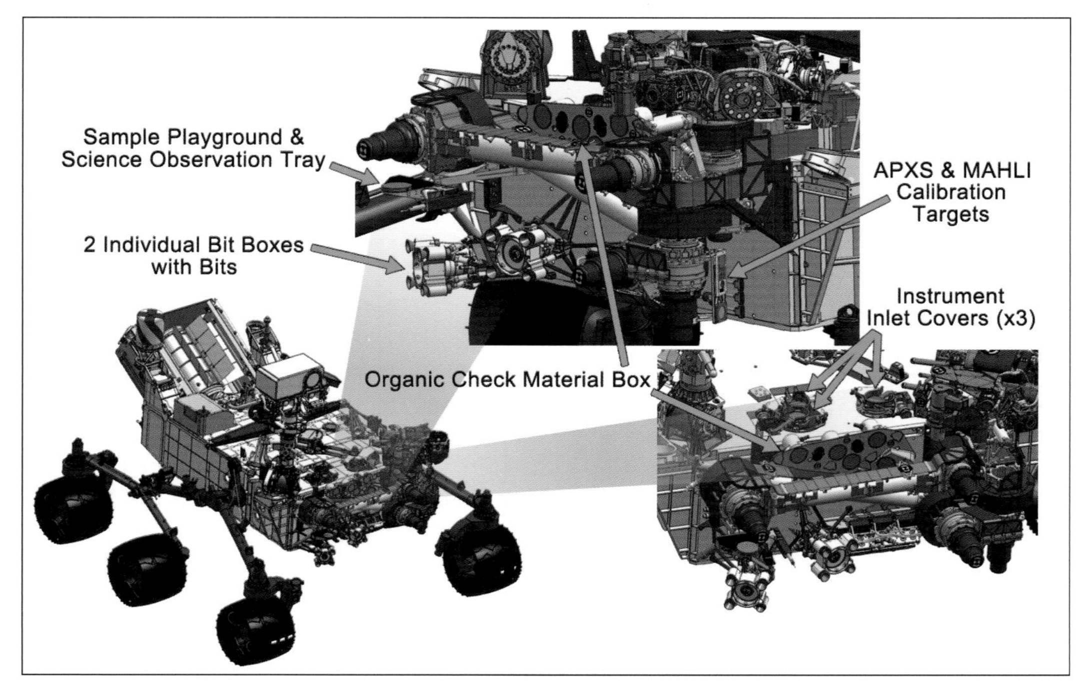

RIGHT Curiosity carries a range of equipment attached to the main body of the lander for providing replacement equipment such as drill bits or calibration targets. *(JPL)*

spectroscopy, the main electronics are housed within the rover WEB, and measurements are taken by the sensor head no more than 0.75in (1.9cm) from the sample. This is much closer than possible with the MER rover instruments and allows for faster data collection. The X-ray emissions can be read for anything from 15 minutes to three hours, after which the data is moved to the electronics. The 32kb data set will include 13 consecutively acquired spectra with detector chips kept below 32°F (0°C) by a Peltier cooler, and this allows the APXS to take readings during the day, unlike APXS instruments on the MER rovers, which were largely restricted to night-time readings.

Curiosity's APXS also has the ability to measure heavy elements such as iron due to its greater sensitivity. At 700 micrograms (60 millicuries) of curium, this instrument has twice the quantity carried by the MER rovers. Because curium-244 has a half-life of 18.1 years it is particularly suitable for long-duration missions. Spectra of about ten seconds will be used to position the instrument with great precision, and a measurement for detecting elemental abundances above 0.5% can be read in ten minutes. To get measurements in the parts-per-million scale requires scans of at least three hours. When in physical contact, the sampled area is 0.7in (1.7cm) in diameter, but stand-off readings can give a greater peripheral sampling area, albeit with a lower intensity.

Essentially, compared to the MER rovers Curiosity's APXS has a threefold increase in sensitivity from a stronger X-ray source, detecting elements with larger proton quantities. Moreover, a new technique that benefits greatly from the added X-ray intensity has been developed at the University of Guelph, Ontario, Canada. Known as the scatter peak technique, it disassociates oxygen, invisible to X-rays, and can help detect water in minerals such as those found in salty subsurface soils, as was the case with Spirit.

This much more capable instrument will permit seasoned experimental techniques for measurement of elemental abundances to an accuracy of about 10%. Although calibrated pre-flight against known minerals, the instrument will periodically be recalibrated against a known basaltic rock surrounded by a nickel plate located on the rover's observation tray. The greater flexibility afforded by carrying the APXS sensor head on the RA turret allows the instrument to check out processed samples on the tray and also to examine soil freshly exposed by the rover's wheels.

Measuring the mineral content and relative abundances in rocks and soil is a key to understanding the past and present environment of Mars. Another 'first' for space exploration on the Red Planet is the Chemistry and Mineralogy Experiment – CheMin – designed to analyse powdered rock and soil samples gathered by the robotic arm. Using X-ray diffraction, supplemented and enhanced by X-ray fluorescence, CheMin can determine to a fine degree the relative compositional abundances by measuring the ratio of particular elements in the sample.

Because minerals are crystalline where atoms

ABOVE Taken in the Spacecraft Assembly Facility at JPL during testing on 3 June 2011, the robotic arm reveals a handful of tools on the turret. To the left is the percussive drill while the MAHLI device is farther across to the right around from the Dust Removal Tool on the front edge. *(JPL)*

ABOVE Curiosity gets down and dirty with this robotic arm and a turret-full of contact equipment. *(JPL)*

are arranged in an orderly structure, an X-ray beam fired at a sample will be scattered at the atomic level in a predictable arrangement of angles according to the spacing between atomic planes in the crystals. This taps into the signature of the mineral and provides information on its type and abundance. This X-ray diffraction method allows scientists access to information about the structure and composition of the sample. The CheMin instrument is installed within the main body of the rover and it is there that the analysis will be made, from where the data will be delivered to the communications system for transmittal to Earth.

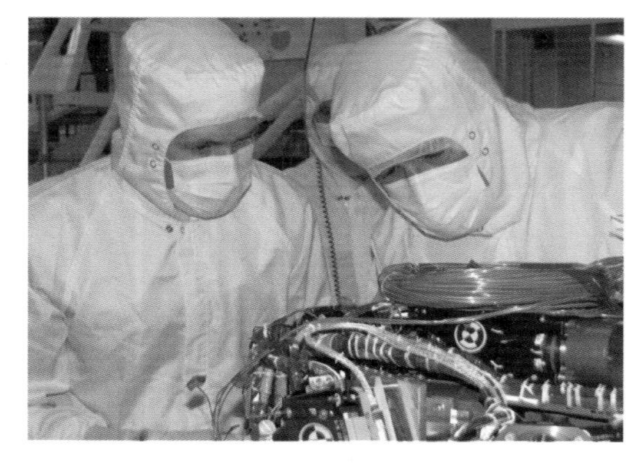

RIGHT Graduate student Nicholas Boyd (left) and principal investigator Ralf Gellert from the University of Guelph, Ontario, Canada, prepare the APXS for installation on the arm turret. *(JPL)*

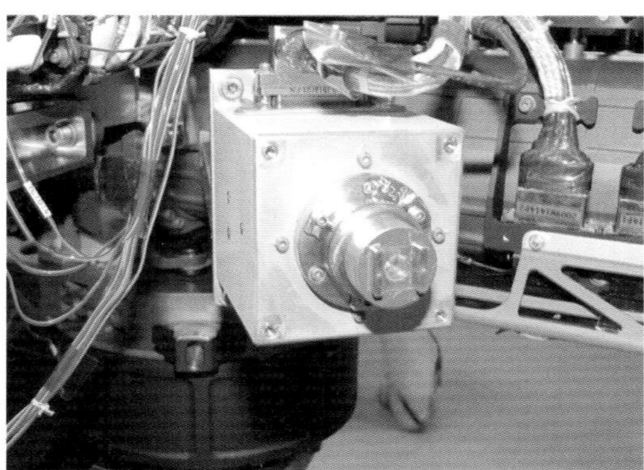

RIGHT About 3in tall, the APXS sensor head is placed on the turret, 18 February 2012. *(JPL)*

Getting the samples into the CheMin instrument is the job of the Sample Acquisition, Sample Processing and Handling (SA/SPaH) system (see later), but it is equipped to analyse as many as 74 samples, each of 27 sample cells being used up to three times. The work is complex and long, each sample analysis taking up to ten hours across two or more Martian nights. CheMin uses a cobalt X-ray source, a transmission sample cell and an energy-discriminating X-ray-sensitive CCD to produce two-dimensional X-ray diffraction patterns and energy-dispersive histograms from powered samples. The raw CCD frames are processed on board to reduce the amount of data generated prior to transmission.

The device itself consists of a funnel, a sample wheel carrying 27 reusable sample cells and five permanent reference standards, and incorporates a sump where used material is dumped. The process begins when material is brought to the top of the rover deck by the SA/

LEFT Remote sensing and contact instruments are caught on display as Curiosity is assembled. *(JPL)*

BELOW The 34mm MastCam camera took this view of the rover's arm-mounted MAHLI camera on Sol 30, revealing a thin coating of dust. The Dust Removal Tool (DRT) is to the right. *(JPL)*

SPaH devices at the end of the robotic arm. Drilled powders or scooped up samples are delivered to a funnel capable of accepting a maximum 65mm^3 of material and vibrated by a piezoelectric device to shake it down. The funnel has a 1mm mesh screen to prevent larger grains falling through, but any that are larger will remain there for the rest of the mission. In fact, particles will have been pre-screened first to 1mm and then to 150μm in the CHIMRA sorting chamber (see later). When not in use the funnel has a protective cover.

BELOW Concerned about the possible build-up of dust during landing, this image taken on 8 September 2012 reveals a clean sensor head for the APXS instrument. *(JPL)*

LEFT Seen in the centre of the picture with its LED lights on, the MAHLI instrument was checked out on 12 September 2012, its dust cover removed. *(JPL)*

ABOVE A technician secures one of two finger-like arms constituting the Rover Environmental Monitoring Station, which will study atmospheric characteristics in Gale Crater. *(JPL)*

Under normal conditions only particles smaller than 150μm in size find their way down the funnel. Grains between that size and 1mm will pass to the upper reservoir of the sample cell and remain there until the cell is inverted and the contents fall into the sump. The design requirement paid particular attention to potential contamination from one sample to the next, and the entire system was built to prevent 95% of material from one sample affecting results from the next. Used samples are removed by the device inverting and then vibrating the cell to shake out the contents. Although some samples will require analytical time of just a few hours, most will require ten hours, and conducting this process at night achieves the desired low temperatures through ambient processes.

The sampling device at the bottom of the funnel is a wheel comprising 32 cells, of which five are preloaded on Earth for control analogues, fixed in pairs like tuning forks. The funnel is at the top of the wheel, which rotates to position an appropriate pre-selected cell directly beneath the funnel outlet into which the contents are dropped. Each sample cell in the tuning-fork pair has circular windows either side, with 13 cell windows covered in Kapton and 14 in Mylar. Kapton is durable under strong vibration and is not susceptible to acid attack, unlike Mylar which has the advantage of a flat diffraction background but is liable to destruction in the presence of strong acids. The remaining five cells contain the preloaded reference materials.

Each disc-shaped cell in a pair has a diameter of 8mm and a thickness of 175μm, with the windows 6μm thick for Mylar and a little thicker for Kapton. A collimated 50μm-diameter X-ray beam is generated by firing high-energy electrons at a target of cobalt. The sample is positioned between the beam on one side and detector on the other. The X-rays penetrate the window at right angles to the wheel and stimulate the sample. A normal excitation vibration of 2.15kHz is used but larger amplitudes can be introduced during the process to prevent domination of the results by denser or slightly larger grains than others.

The detector in CheMin has a 600 x 600 E2V CCD-224 frame transfer imager to collect array pixels of 40 x 40μm^2 and is a modernised version of the E2V CCD-22 designed and built especially for X-ray telescopes used by astronomers. It must be kept at a temperature below –76°F (–60°C). The larger size of the individual pixels allows a larger number of X-ray photons to dissipate their charge inside a single pixel and this improves the collection of high-energy X-rays. The detector reads secondary X-rays excited by the primaries and this provides

BELOW Development of the robotic arm was one of the demanding challenges for engineers, the 75lb turret of instruments needing lots of muscle to move and hold it in place. Here, weights simulate the mass of the turret during tests on 12 June 2009. *(JPL)*

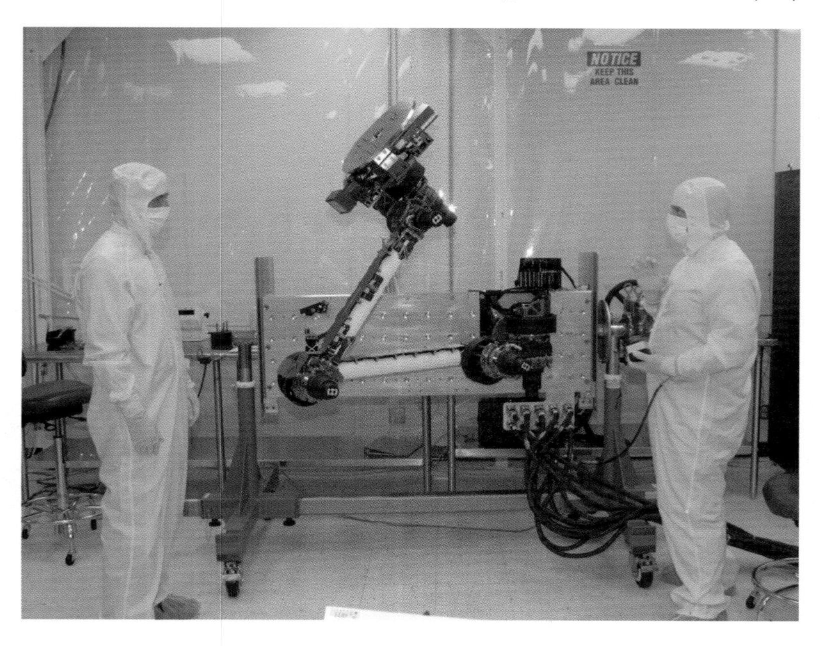

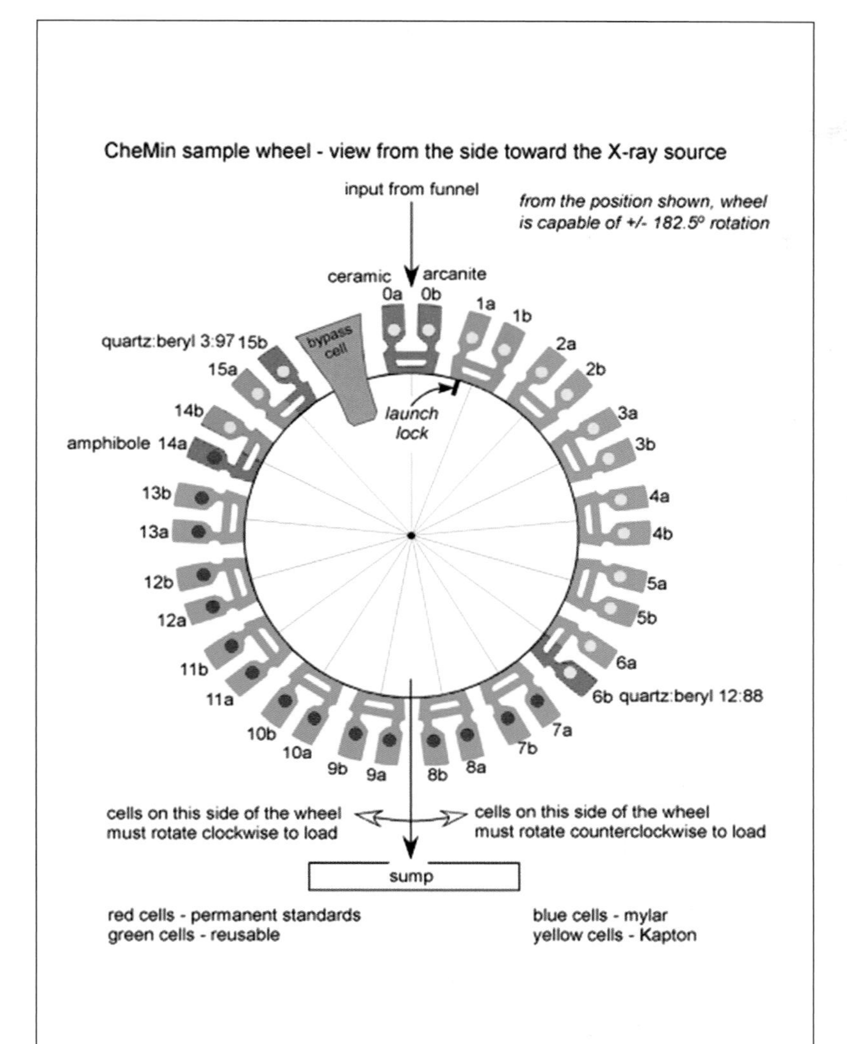

LEFT The Chemistry and Mineralogy (CheMin) sample wheel has 27 reusable cells to which tiny quantities of soil are delivered in sequence for analysis. *(JPL)*

ABOVE Sample pairs are arranged like tuning forks within the compact CheMin wheel so that appropriate rotation will place the assigned receptacle directly beneath the feed funnel at the top. *(JPL)*

BELOW LEFT AND RIGHT The CheMin funnel located on the top of the rover upper deck in closed and open positions, seen from a distance of about 8in. *(JPL)*

ABOVE Weighing 88lb, the Sample Analysis at Mars (SAM) suite of instruments is one of the more complex and sophisticated experiments yet sent to Mars, tasked with examining samples for information about the chemicals that could indicate past environments amenable to life. *(JPL)*

BELOW The three principal SAM experiments include a Quadrupole Mass Spectrometer, a Gas Chromatograph instrument and a Tunable Laser Spectrometer incorporated within a compact box structure. *(JPL)*

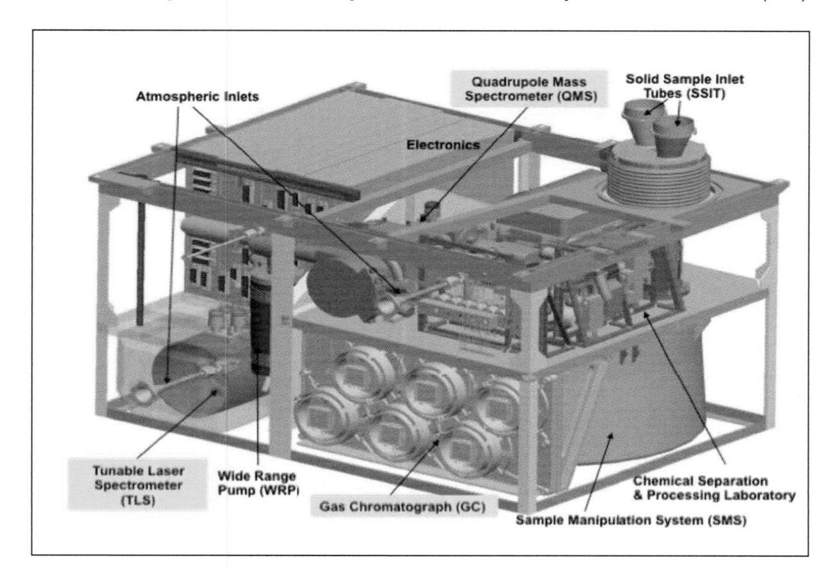

the fluorescence data for detecting elements higher than sodium in the periodic table.

CheMin is not the only instrument inside the rover to use materials delivered by the toolkit on Curiosity's robotic arm. Designed as a group of three separate investigations, the Sample Analysis at Mars (SAM) experiment incorporates a Quadrupole Mass Spectrometer (QMS), a Gas Chromatograph (GC) and a Tunable Laser Spectrometer (TLS). All three are installed within the SAM container in the forward part of the rover's interior structure. It is the largest suite of instruments by volume but contains instrumentation that on Earth would require a very large section of a standard laboratory!

The purpose of the SAM experiments is to search for carbon-based compounds, believed to be fundamental in the building blocks of life, and it will examine the gases from the Martian atmosphere and pyrolyse powdered rock and soil samples to study those gases too. The robotic arm will deliver physical samples but ambient gaseous samples will be drawn from atmospheric gases that valves and pumps draw in through filtered inlets ports on the side of Curiosity.

The principal goals of SAM are to conduct a survey of carbon compounds and determine their possible mechanisms of formation and destruction, search for organic compounds important for prebiotic and biotic life, reveal the chemical and isotopic elements important for life, seek out trace atmospheric components for evidence of interaction with the soil, and measure the noble gases and light element isotopes to improve understanding of the evolution of the atmosphere and the climate on Mars.

Scientists want to use these five tasks to answer three fundamental questions: what does the presence (or lack) of carbon compounds at or near the surface say for its potential for life; what do the results from the search for lighter elements say about their interaction with solids and gases to reveal potential habitability for life past or present; and while accepting that the climate of Mars was different in the past, what does that say about the potential for life having developed?

SAM is a complicated package that has placed demands on electrical power for efficiency and effectiveness, drawing on a limited power supply. It contains two ovens capable of heating powered materials to 1,800°F (1,000°C) yet draws only 40W of power. But the operation of the three science investigations and the collection and transfer of data requires a complex design requiring about 2,000ft (610m) of wiring. Stimulated by the frustration of data on the Viking landers, the search for organic materials has driven the technology in SAM to provide a system on Curiosity that can seek out a greater range and variety than any other equipment sent to Mars.

The Quadrupole Mass Spectrometer is a familiar piece of equipment, carried by SAM to measure gases by molecular weight and identify them by the electrical charge of their ionised states. In the search for organics, it will identify nitrogen, phosphorous, sulphur, oxygen, hydrogen and carbon. The Gas Chromatograph separates the different gases in a given mixture, detects organic compounds emerging from a capillary column and delivers them to the QMS. The Tunable Laser Spectrometer measures the absorption of light at particular wavelengths to measure the amounts of methane, carbon dioxide and water vapour and to identify the different isotopes. The relative ratios are useful indicators of the processes at work in planetary environments and can tell much about past atmospheres.

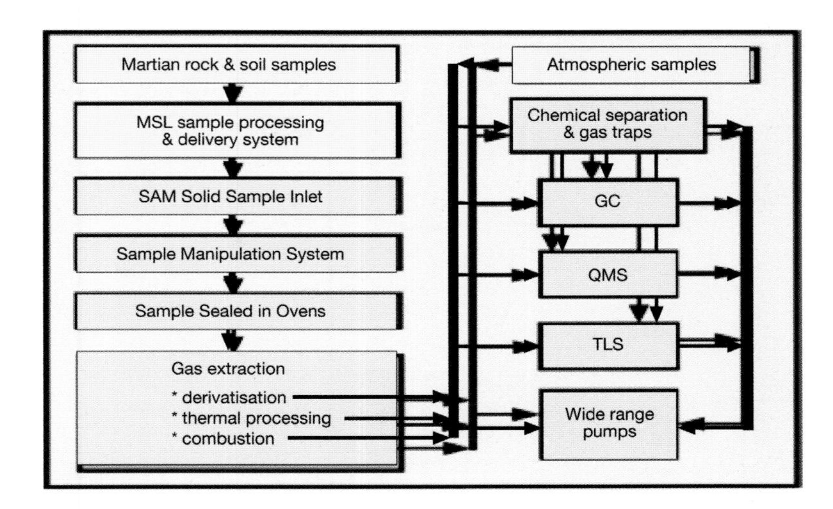

ABOVE The flow path for the three separate experiments incorporates physical and gaseous materials taken from the surface and the atmosphere. *(JPL)*

All solid materials for analysis in the SAM go through the same delivery and preparation process before being sent to the three science instruments. It includes a sample manipulation system where finely sieved materials are delivered to one of 74 sample cups, 59 of which are made of a quartz so that samples can be inserted into an oven and thermally processed for the release of volatiles. Each cup has a volume of 0.78cm^3 and after use it can be baked free of any contaminants and used again. Of the remaining cups, six carry solids for calibration purposes and nine are available for a process known as derivatisation for extracting gases from samples.

Derivatisation is a technique for defining more precisely larger and more reactive molecules in a region discovered to have a higher abundance of organics. It is a process that uses these special cups in the SAM to mix powdered material with a solvent mix and a chemical that will react with the sample at a lower oven temperature, producing quicker reactions. The nine cups are sealed with a foil cap until punctured by the sample manipulation system, whereupon the powdered material is introduced. In this way the reactions produce a volatile compound that will be separated out in the Gas Chromatograph. This flexibility in analytical methods is beyond the limited capabilities of the science suite carried by

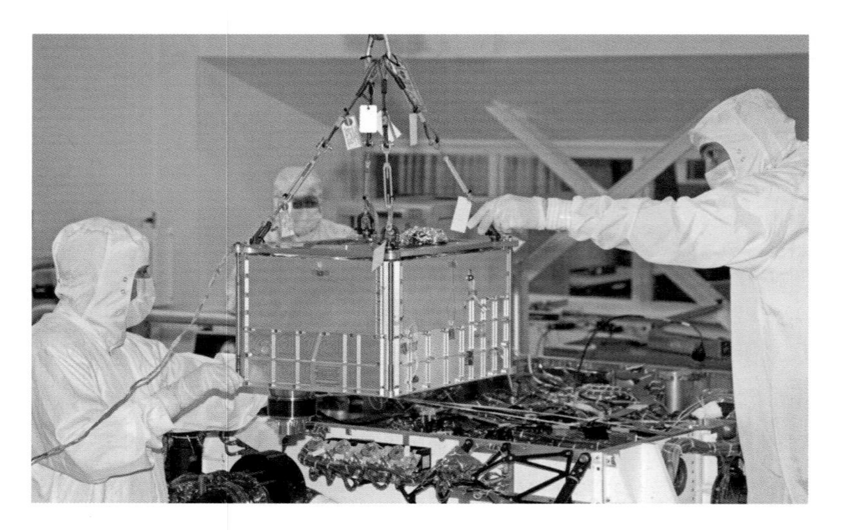

ABOVE A mechanical hoist manoeuvres the delicate SAM box through the top deck of the rover. *(JPL)*

the Viking landers, which failed to identify any organic compounds.

Prior to Curiosity the mantra for Mars programme planning was 'follow the water', implying that if water could be found beneath the surface it was possible that life could once have evolved on the planet and that these places would be the best when searching for evidence of biological activity. For much of the time after Viking the Mars missions were planned around that goal. Viking blew apart the seemingly narrow gap between chemistry and biology and showed that the presence of water

RIGHT SAM is lowered gently into the Warm Electronics Box for a controlled thermal environment. *(JPL)*

in itself is not enough to promote the notion of proximal life or evidence that it once existed there. Now, the mantra has shifted to 'follow the carbon', which is why the Gale Crater site is well chosen for that search.

Some of the results from the two Viking sites may have been distorted by an underlying causal effect of perchlorates having masked the prior existence of organic compounds in soils analysed by the spacecraft. The process of derivatisation in Curiosity could yet reveal substantive evidence for the existence of organic materials subsequently veiled by chemical action. Moreover, the fact that the Viking landers could only sample an area within a narrow arc directly in front of the remote sampling arm greatly limited a proper survey of the site. For the first time on Mars, an advanced roving vehicle can conduct highly detailed surveys, measurements and analyses across a wide part of the surface extending to many kilometres from the touchdown site.

A key to the operation of SAM's three scientific experiments is the chemical separation and processing laboratory with its pumps, pipes and tubing as well as gas reservoirs, pressure monitors and temperature monitors. It has 52 microvalves for directing the flow of gas through the system as well as two vacuum pumps designed to rotate at more than 1,666 revolutions per second for maintaining pressure within the instruments. There are checks and balances too for scientists to remotely manipulate sequences to verify or corroborate findings from specific tests. For instance, should derivatisation, brought on by a prior survey for organics, indicate a presence of molecules, there are five encapsulated bricks of organic material with which to verify a finding.

These silicon-dioxide bricks are impregnated with synthetic fluorine organic chemicals not found on Earth and not expected to be found as part of a natural process. The control method will be to use the same drilling and processing system as that which initially delivered the materials showing organic molecules to take a portion of the powdered brick and subject it to the same process. If organic molecules of the same type as those recorded earlier are found in material from the control brick it will indicate the possibility

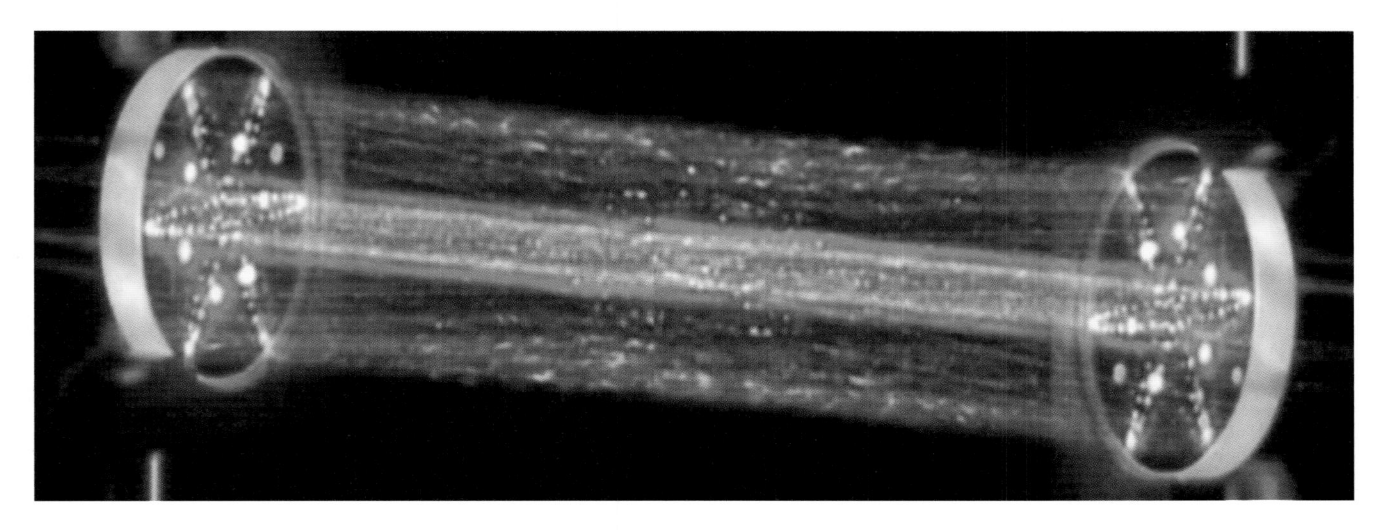

of contamination from Earth. If only fluorine markers are found it would raise the possibility of organic activity, because in that case the previously detected molecules could only have come from the surface of Mars.

Through this scrupulous process the SAM instruments could discriminate between organic molecules that have particular significance in the evolution of life, such as amino acids, the building blocks of proteins. Methane, produced by living organisms, is believed to exist in very small quantities in the atmosphere of Mars, indicating it may once have had life that produced this gas. The Tunable Laser Spectrometer will be able to detect methane, monitor variations in concentration and help determine whether the methane is produced by biological activity or by normal geological processes – also a possibility. The SAM suite was developed by NASA's Goddard Space Flight Center, with Paul Mahaffy as the principal investigator. The Gas Chromatograph was developed by France's national space agency (CNES) and the laser spectrometer was developed by JPL.

None of the science designed into the SAM

ABOVE A demonstration of the Tunable Laser Spectrometer displays the scattered laser beam bouncing at each mirror end of a tube which on Mars could be filled with atmospheric gases. This can measure the different quantities of carbon dioxide, methane and water vapour in the atmosphere. *(JPL)*

LEFT SAM inlets on top of the rover upper deck into which samples of Martian soil are placed for transport to the various instruments. *(JPL)*

suite and its three principal instruments, or to the CheMin, would be possible without the means to obtain and deliver samples to their receptacles. To accomplish that task, Curiosity has four specific pieces of equipment on the turret at the end of the robotic arm: the Sample Acquisition/Sample Processing and Handling (SA/SPaH) subsystem, the Powder Acquisition Drill System (PADS), the Dust Removal Tool (DRT) and the Collection and Handling for In-Situ Martian Rock Analysis (CHIMRA). The SA/SPaH is a collective grouping of the robotic arm and its turret devices, the drill and drill bit boxes, a solid scoop, a sample processing subsystem, organic check material and an observation tray.

Starting with the method of extracting powdered rock for the SAM and CheMin instruments, the PADS will obtain samples to a maximum depth of 2in (5cm) from a drilled hole 0.6in (1.6cm) in diameter. It consists of a drill and an internal auger up which the powdered material is transported to the CHIMRA, which has the role of sieving and apportioning samples. In a typical sequence, the drill is placed in contact with the surface – a telemetry switch ensuring that contact has been made – while being preloaded on to the target by the robotic arm to reduce the likelihood of it moving during drilling. Samples are obtained by either rotating or hammering, or by a combination of both, after which it is disengaged from the hole and placed in a stowed position.

As a result of tests the drill was designed as a rotary percussion type, showing much less bit wear than a rotary drag type, especially on rock with a high compressive strength. The drill has been designed to provide 81 samples to the experiments on board and two spare bit assemblies are provided in a dedicated bit box assembly attached to the front panel of the rover. An innovative design involving rotation, translation and movement of the drill chuck allows the active drill to conduct its own extraction of replacement bit. Moreover, should a drill bit become stuck in a rock, threatening to anchor the rover to a specific spot, the drill can be commanded to abandon the bit and replace it, thus freeing Curiosity. The JPL engineering team tasked with designing the PADS built 12 test drills subjected to rigorous trials and evaluation to produce the definitive device.

The drill is an assembly of seven complex elements. From the front end these include: the bit assembly to cut rock and collect the sample; a chuck mechanism to engage and release new and worn bits; a spindle mechanism to rotate the bit; a percussion mechanism to generate hammer blows to break the rock; a translation mechanism to create linear motion, which incorporates a force sensor; a passive contact sensor to secure the drill's position on the rock surface; and a flex for electrical power and command signals to operate the device.

The bit assembly is a standard ⁵⁄₈in (16mm) commercial hammer drill with a shank turned down and machined with deep flutes for aggressive cutting. Surrounding the shank is a thick-walled maraging steel collection tube that allows the powdered sample to be augered up into the sample chamber. The 0.9lb (0.4kg) drill percussion mechanism and hammer operates at 1,800 strikes per minute with variable impact options ranging from 0.05 to 0.8 Joules. The translation mechanism maintains a force of 120N on the bit during operation, generating the extraction force to remove the bit from the rock or for attaching the drill to a fresh bit in the bit-box.

Dust and loose material that could clog filters or disturb the efficient running for the sample distribution equipment is removed by the Dust Removal Tool held on the arm turret. This will be used for cleaning all surfaces of extraneous material and consists of a stainless steel wire brush driven by a single actuator and capable of cleaning an area with a diameter of 2in (4.5cm) without moving the arm. The DRT is also used for cleaning off the observation tray. Comprising a circular disc 3in (7.5cm) across, the tray is used for contact analysis of small samples by the APXS or MAHLI on the turret and not for samples heading for the CheMin or SAM instruments inside the rover.

The CHIMRA device incorporates a clamshell-shaped scoop just 1.6in (4cm) wide, which is used to collect small fragments from the surface of Mars. Another section of the CHIMRA has a series of chambers and labyrinths designed for sorting, sieving and apportioning the various samples retrieved by the drill and the scoop. Two sieve screen

RIGHT The Collection and Handling for In-Situ Martian Rock Analysis (CHIMRA) is attached to the turret at the end of the robotic arm, delivered either by the drill through the top transfer tube or from the scoop at the bottom. *(JPL)*

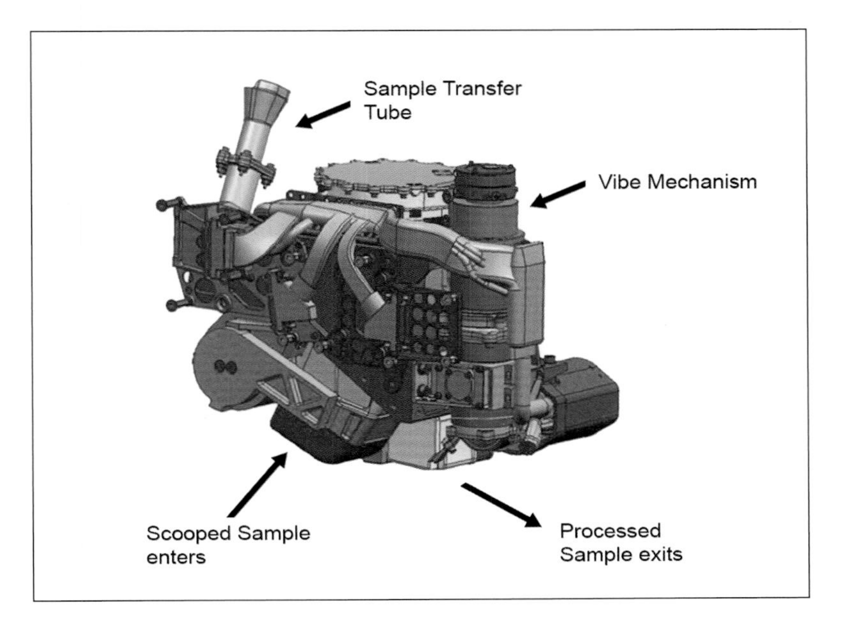

sizes are available for filtering material more than 0.04in (1mm) or 0.006in (150 microns) in size. Volumetric capacities are graded to 45–130mm^3 for the larger size and 45–65mm^3 for the smaller size. The dispensing tubes are known as 'thwacks', for respective grain sizes.

The device can be vibrated at 3–10gs to help differentiate portion size, and gravity is used for the contents to fall from the sieve into the appropriate receptacle by rotating that particular outlet to the inverted position. The vibrating action only requires a force of 150N because the CHIMRA device on the turret weighs only 18lb (8kg). The motor drives a central shaft and eccentric mass through a helical coupler at a speed of 4,000–6,000rpm. Three inlet ports are provided on the rover itself, two for the three SAM instruments and one for CheMin, and all have motorised covers closed tight when not in use.

All the instruments, experiments and sampling devices installed on the Remote Sensor Mast, the robotic arm and the interior of the rover support physical examination of materials on the surface. The Rover Environmental Monitoring Station (REMS) was installed on the RSM for recording weather data and for assembling information about the changes to the environment in Gale Crater over at least a full Martian year. Viking was the first to take direct measurements, in situ, of the weather on Mars, and Curiosity will continue the tradition set by Pathfinder and the two MER rovers. In fact, since January 2003 not a single day has gone by without a daily weather reading from one location or another occupied by a rover or a lander.

The REMS device is supplied by Spain and records wind speed and direction, air pressure, relative humidity, air and ground temperature and levels of ultraviolet radiation. The equipment is designed to take a reading lasting five minutes every hour for the full duration of the mission. Wind, temperature and humidity is measured by two booms attached halfway up the RSM and positioned 120° apart, one (boom 1) toward the front of the rover and the other (boom 2) to the side so that no part of the structure can shield the sensors from the effects of the weather.

Both booms carry a single sensor for air temperature and three sensors for three-dimensional modelling of air motion. Boom 1 carries an infrared sensor to get a reading on the ground temperature and boom 2 has the humidity sensor pointing downward inside the cover. The infrared sensor is situated on the

BELOW An internal view of the CHIMRA device shows, in pink, the route followed by scooped samples. *(JPL)*

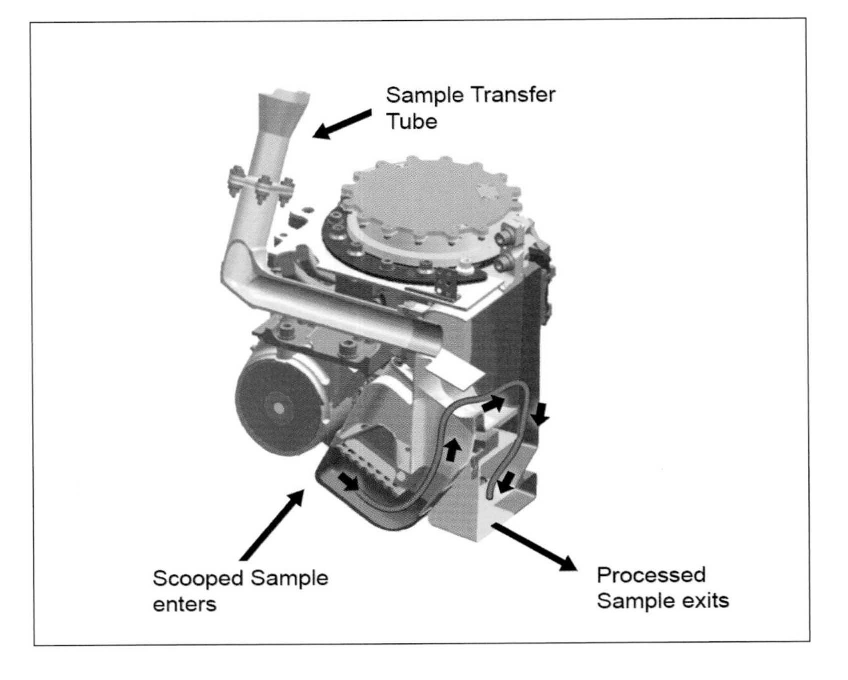

RIGHT The CHIMRA scoop in contact with a test sample. *(JPL)*

deck of the rover measuring six wavelengths that are also read from Mars Reconnaissance Orbiter, but the surface measurements are a first for landed spacecraft.

A full understanding of the atmosphere of Mars is essential if humans are ever going to go there, build science bases from which to foray on expeditions across the surface and establish the first colonies. A lot is known about the dynamics of the Martian weather but one area where scientists need to know more is the amount of radiation reaching the surface and at what levels. The surface of Mars itself is bombarded with radiation from the Sun, and from galactic sources elsewhere in the universe, but a full map of particle population density is not available. This information is important for the detail it can provide on the physics of the planet as well as the hazards faced by people.

When the Mars Science Laboratory project was being developed, NASA had made a commitment to develop the technologies for carrying humans back to the Moon and on to Mars. NASA scientists and engineers planning the next generation of spacecraft capable of carrying humans deep into the solar system were keen to have MSL measure these radiation levels all the way from Earth to the Red Planet and at frequent intervals across the surface. Uniquely, MSL got some money from those groups at NASA working on such planning and that helped fund the Radiation Assessment Detector (RAD).

The RAD is a 3.8lb (1.7kg) instrument incorporating a wide-angle telescope angled upward inside the forward portion of the rover on the left side. It can detect neutrons and gamma rays whether from the atmosphere or from subsurface materials. Energetic particles from the Sun are potentially harmful to life and the surface of Mars is a particularly bleak place for experiencing these surges. Solar storms, sometimes unpredictable, can blast the Red Planet with very harmful levels of radiation.

Mars has a very thin atmosphere, less than 1% that of the Earth, and no global magnetic

BELOW Fine sand is filtered through the CHIMRA device on the first 'decontamination' exercise, passing through a sieve with 0.006in mesh as viewed from Curiosity's MastCam. *(JPL)*

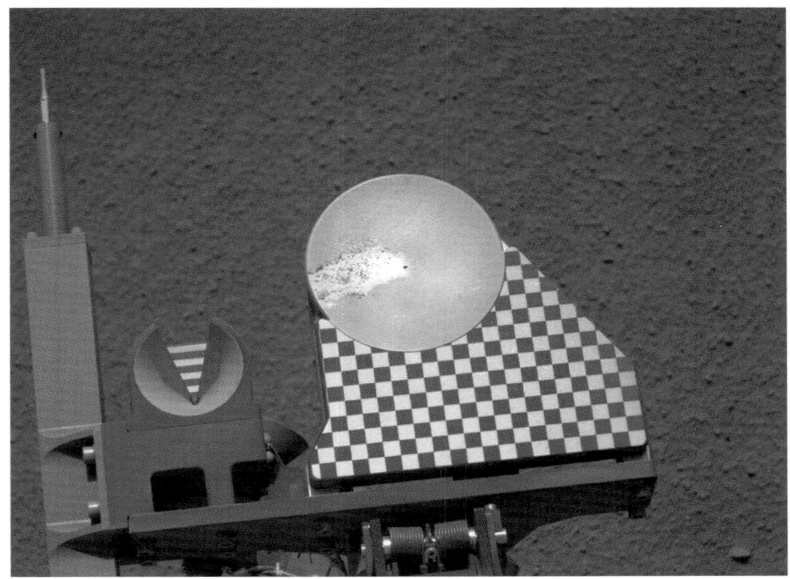

ABOVE LEFT Residue of particles larger than 0.006in left after the scoop has deposited fine dust to the CHIMRA experiment on 10 October 2012, Sol 64. *(JPL)*

ABOVE A sample of soil was delivered to Curiosity's 3in-diameter observation tray for the first time on 16 October 2012, Sol 70 since landing. *(JPL)*

field to protect the surface from harmful particles that are deflected from the surface of our own planet. If seeking signs of microbial life, future missions would benefit from RAD measurements in knowing how far they had to dig beneath the surface to find a zone where harmful particles do not penetrate. The RAD is a collaborative project between Southwest Research Institute in Boulder and San Antonio, NASA's Exploration Systems Mission Directorate and the German national aerospace research centre.

Although there is much evidence that water once flowed on Mars, the atmosphere is now too thin and the pressure too low for free-flowing water today. But surviving remnants of that wetter epoch may be found in hydrated minerals, hydroxyl ions bound into their crystalline structures that could lie just beneath the surface. To search for such remnants, in cooperation with NASA, Russia has supplied an instrument to scan the surface over which Curiosity rolls by bombarding the ground with neutrons to measure how they are scattered in a search for hydrogen to a depth of 20in (50cm). Known as the Dynamic Albedo of Neutrons (DAN), the instrument can be useful in working with other science tools and can even trace the subsurface root of rocks visible to the camera above the surface.

DAN works on the principle that when high-energy neutrons collide with hydrogen atoms they are decreased in energy and scattered.

BELOW The location of the Dynamic Albedo of Neutrons instrument incorporates a neutron generator on the right side of the rear deck as shown, with the detectors on the opposite side in the same relative position. *(JPL)*

Measurement of the reflected atoms can determine the small quantity slowed by collisions which is proportional to the amount of hydrogen encountered. Similar scans of the surface from orbit were made by a Russian instrument of Mars Odyssey, apparently detecting underground water ice just beneath the surface close to the polar regions. This instrument passively observed the effects of galactic cosmic rays striking the surface and inducing the effect, but, while it too can passively observe cosmic ray reactions, DAN can actively bombard the surface beneath the rover to detect water content as low as 0.1% by mass.

DAN is an active/passive spectrometer that measures the time-delay curve of the neutron flux induced by a pulsating 12MeV neutron source. It can be used during traverses, on short stops and when Curiosity is parked. Short duration measurements of less than two minutes can provide a rough estimate of probabilities that hydroxyls exist beneath the surface, but long measurements of around 30 minutes in duration can derive the vertical distribution to an accuracy of 0.1–0.3% by mass. The neutron generator is situated on the right side of the rover towards the rear, about 2.6ft (79cm) above the surface, with two detectors on the opposite side of the rover. Pulses last around one microsecond and can be repeated at a rate of about ten bursts a second. The generator will produce around ten million neutrons with each pulse and the instrument is capable of making ten million pulses over the lifetime of the mission.

Curiosity on Mars

No flight is 'textbook' and without some flaws or heart-stopping moments, but from the time it was launched at 10:02am local time from Cape Canaveral Air Force Station on 26 November 2011, MSL was one of the cleanest missions ever flown. Lift-off came one day into the launch window, a delay caused by a faulty flight-termination battery which would have powered the launch abort system had the rocket experienced a malfunction. But it did not, and the mission quickly settled into its 254-day cruise period between Earth and the Red Planet. Two trajectory correction manoeuvres (TCMs) were made to put the spacecraft on a proper course for

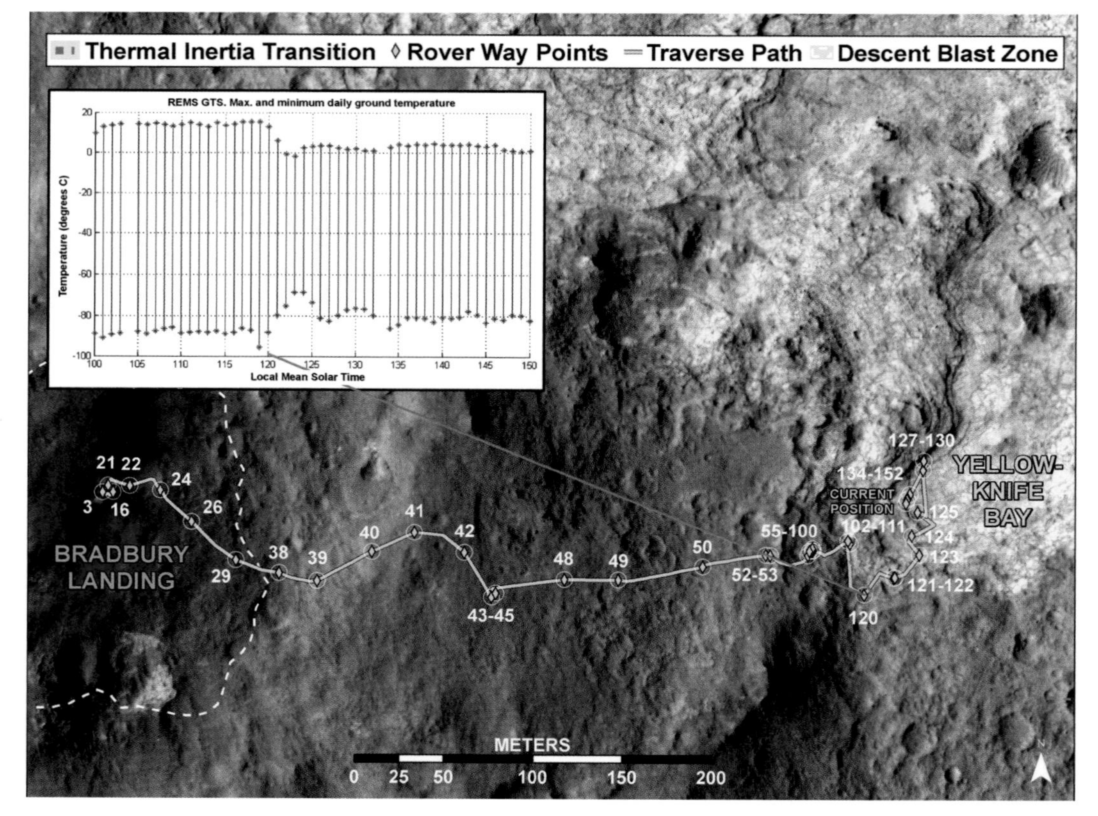

RIGHT The traverse of Curiosity from Bradbury Landing to Yellowknife Bay is traced, with an inset documenting a change in the ground's thermal properties at a different type of terrain. Between Sol 120 and Sol 121 (7–8 December 2012), Curiosity crossed over a terrain boundary into lighter-toned rocks with high thermal inertia values. The green dashed line marks the boundary between the terrain types. The inset shows the range in ground temperature recorded each day by Curiosity. Yellowknife Bay contains sulphate-filled veins and concretions discovered in the Sheepbed Unit, along with much finer-grained sediments. *(JPL)*

Mars; the course set by the Centaur upper stage would miss the planet so that the inert vehicle would not follow MSL into the atmosphere of Mars! It was what engineers call a hybrid trajectory – initially offset but realigned for slicing into the atmosphere of Mars in early August 2012.

The first course correction (TCM-1) took place on 11 January 2012 for thruster firings over a period of 59 minutes for a total velocity change of 12.3mph (5.5m/sec). This effectively shifted to flight path from an initial miss distance of 25,000 miles (40,000km) to a point just 3,000 miles (5,000km) from Mars and by 14 hours in arrival time. TCM-2 on 26 March involved nine minutes of firings, changing velocity by a mere 2mph (0.9m/sec) to place MSL on a Mars intercept course. TCM-3 was conducted on 26 June to refine the entry angle and arrival time, advancing it by 70 seconds, in firings lasting 40 seconds that changed the entry point by 125 miles (200km). TCM-4, the last course correction, took place on 28 July in two brief thruster firings lasting seven seconds, changing speed by 0.4in/sec (1cm/sec) and adjusting the atmosphere entry point by 13 miles (21km), signalling the start of the eight-day Final Approach Phase.

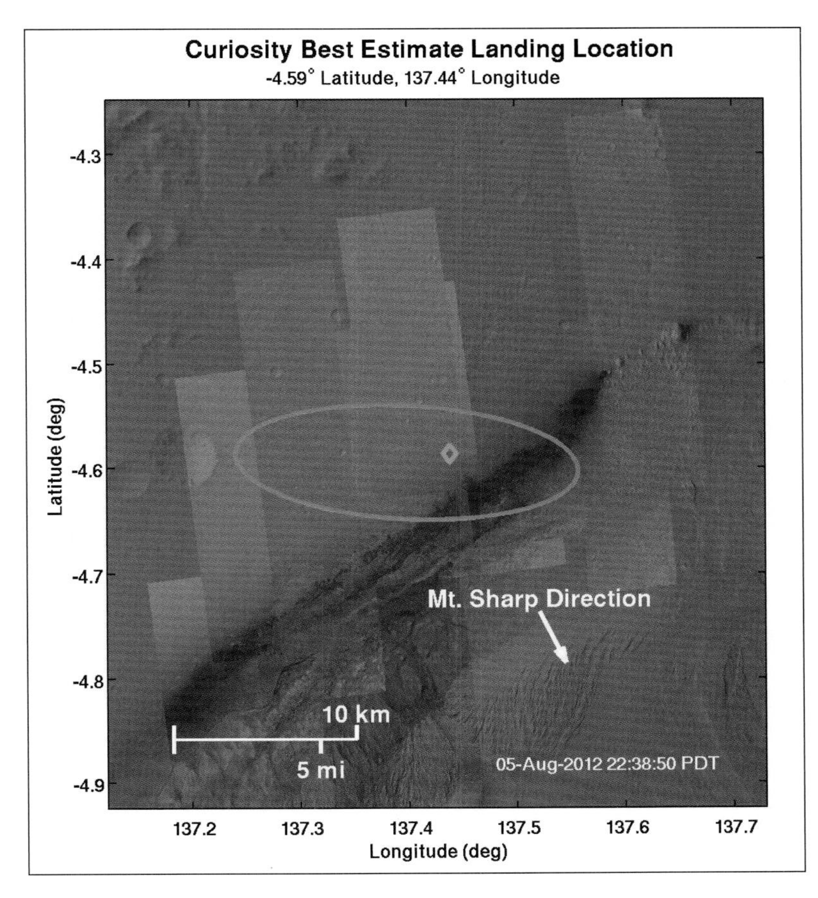

ABOVE Curiosity touched down about 1.4 miles away from the centre of the target ellipse marked in blue, with Mount Sharp away to the bottom right. *(JPL)*

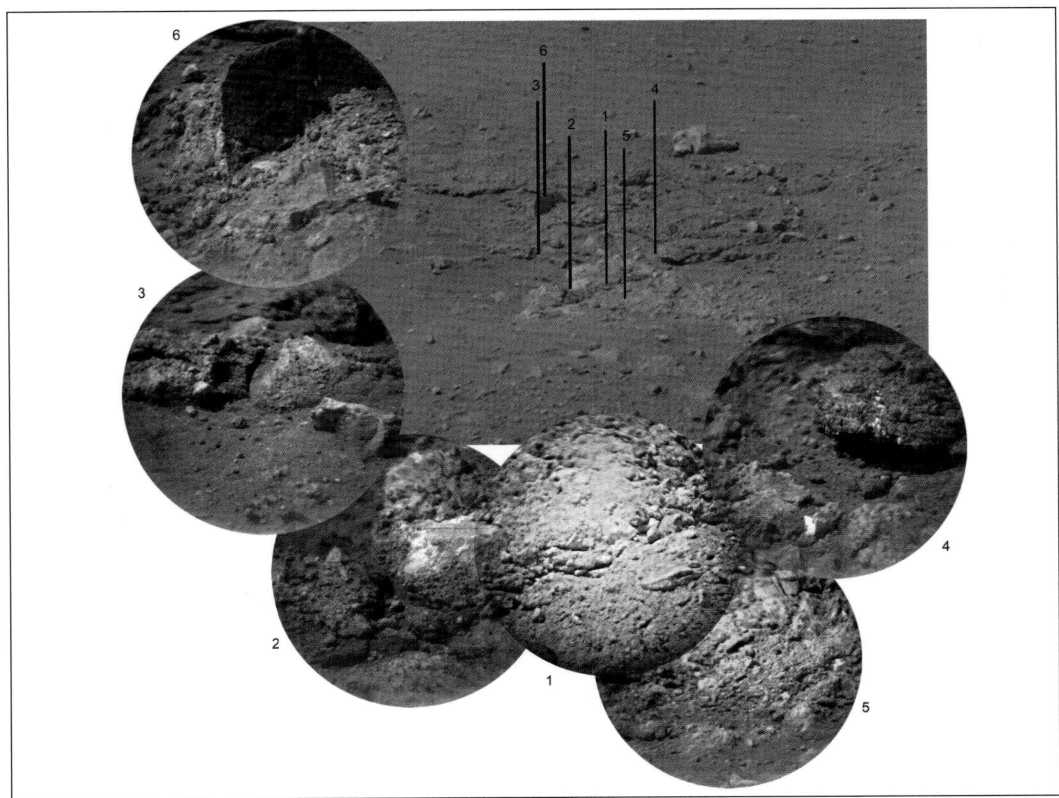

LEFT This mosaic of six images taken by the Remote Micro-Imager on the ChemCam instrument shows marks left by the rocket engines on the Powered Descent Vehicle about 16ft away from Curiosity. *(JPL)*

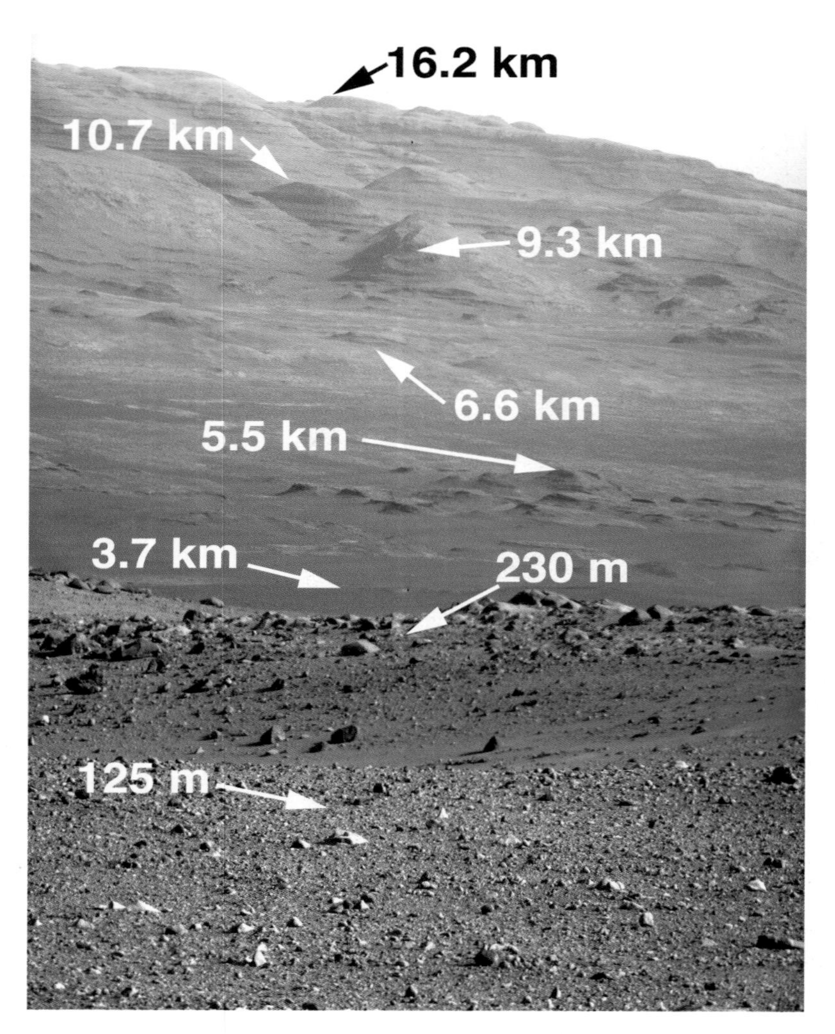

ABOVE An image from the 100mm MastCam camera showing the distance to various elevations on Mount Sharp. *(JPL)*

The precision with which MSL arrived at Mars was crucial for imaging and relay purposes through Odyssey and Mars Reconnaissance Orbiter in their relatively fixed paths around the planet. Too soon or too late arriving for the dash through the atmosphere and coverage would not be possible. One highly demanding but self-imposed challenge was for MRO to take pictures of Curiosity descending on its parachute with a near-90° angle between the flight path of the descending spacecraft and the near-polar orbital plane of MRO. It was achieved to the enormous pride and satisfaction of the team responsible for that acclaimed event, the image being flashed around the world to show the precision and sophisticated position-fixing of man-made spacecraft around other worlds.

Curiosity touched down at 10:17.57pm on 5 August 2012 JPL time (05:17.57am UT time on 6 August; 03:00 pm on Mars) just 1.4 miles (2.28km) from the centre of the planned ellipse, a location of 4.59°S by 137.44°E, close to the foothills of Mount Sharp in the floor of Gale Crater. The touchdown took place at a vertical landing speed of 2.21ft/sec (0.6739m/sec) or 1.5mph (2.4km/h), with a horizontal drift at touchdown of a mere 0.15ft/sec (0.04437m/sec), or 0.1mph (0.16km/h). The rover was on a heading of 112.7°, facing east-south-east. Engineers calculated some 310lb (140.6kg) of

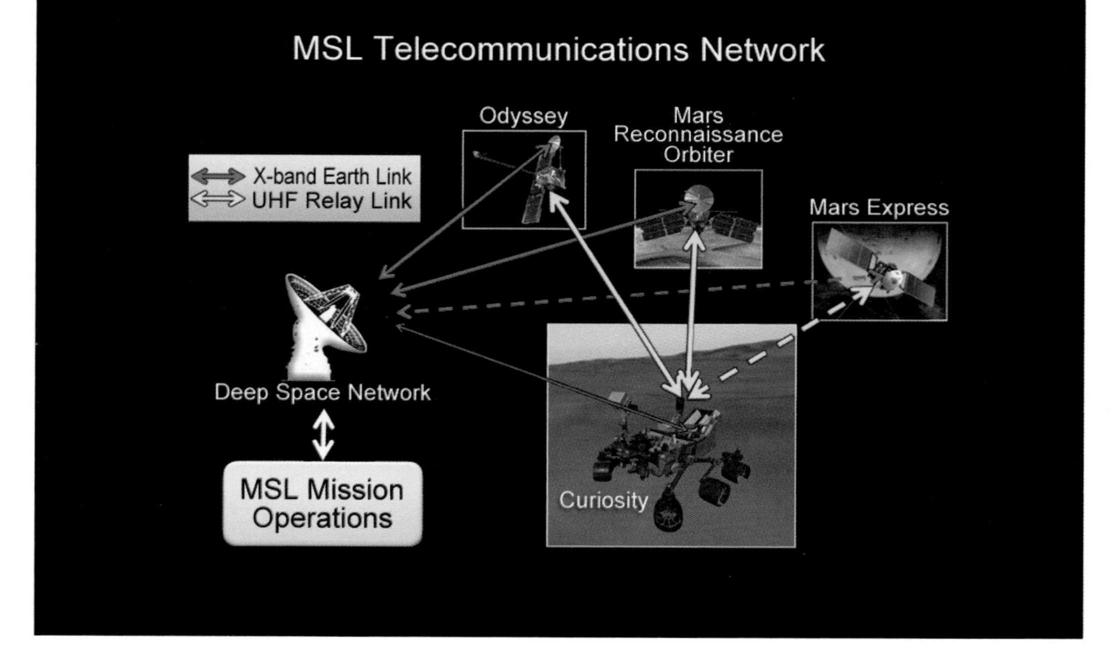

RIGHT The various communication paths for Curiosity between Mars and the Earth and via orbiting spacecraft. *(JPL)*

LEFT The straight lines left in the tracks are Morse code for the letters JPL, with the code on all six wheels being marked as .--- (J), .--. (P) and .-.. (L). *(JPL)*

propellant remained in the Powered Descent Vehicle at fly-away versus an anticipated 203lb (92kg), indicating a conservative descent path to touchdown. The PDV flew away to crash 2,133ft (650m) away from Curiosity.

With the technology worked into the spacecraft and the advanced and sophisticated nature of the entire EDL phase, there was no real surprise that it was the most accurate landing of any unmanned spacecraft anywhere, and a good deal more accurate than many of the Apollo landings on Earth, but there was relief. The site in Gale Crater was renamed Bradbury Landing following the death of Ray Bradbury, whose science fiction writings did so much to entertain and inspire new generations of space explorers. Immediately after landing planning schedules moved from EDL mode to Sol 0 mode, and for several days Curiosity was checked out and all the systems and subsystems tested and brought to life. It would be several weeks before the rover got into its stride.

A major activity began on Sol 3 when the High-Gain Antenna was meticulously bore-sighted on Earth in preparation for a complete software transformation. Over several days the existing R.9.4 EDL package would be replaced with the R10 Surface Operations Software in RCE-A, followed by a similar upload into RCE-B, giving the rover autonomous and hazard avoidance control. R10 had been uplinked to Curiosity in June 2012 during the cruise phase but it had yet to be installed. Taking a cautious approach, engineers first commanded the computer to boot up with the R10 package to determine if it was initialising as anticipated, followed by a full installation in RCE-A and then RCE-B, which took place on Sols 6 and 7. This enabled the full suite of science instruments to be activated, as well as full operations with the robotic arm.

Within two weeks of landing Curiosity had conducted its first ChemCam LIBS laser firing and begun a series of tests and evaluations,

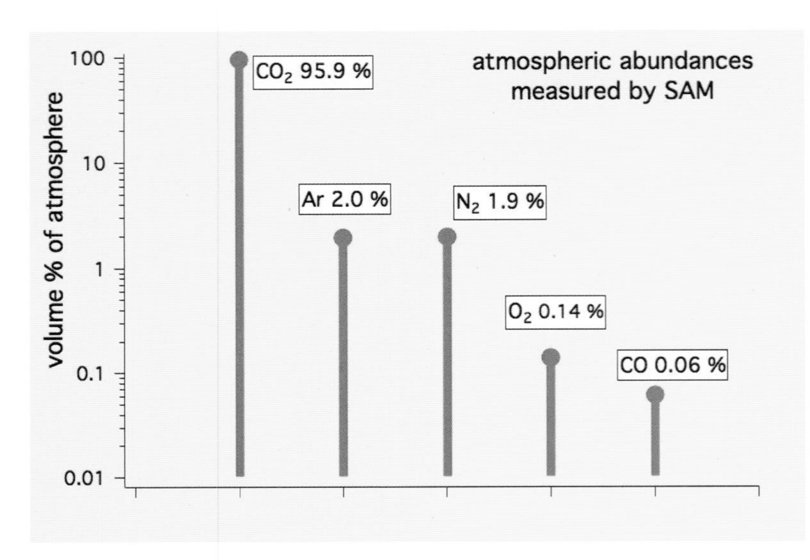

ABOVE The Quadrupole Mass Spectrometer in SAM detected atmospheric gases by volume as depicted on this logarithmic scale, showing a 95.9% abundance of carbon dioxide. *(JPL)*

checking out its operations at the surface. On Sol 15 the mobility system was activated, followed the next day by the first drive, a modest roll of 15ft (4.5m) forward, a 120° turn and a drive back of 8.2ft (2.5m) beginning at 1:30pm local Mars time. This concluded the first part of the Commissioning Activity phase, which now turned to the second part involving tests with the arm, an extensive series of trials starting on Sol 30.

By early September 2012 Curiosity was ready to begin traversing to its initial sampling site, a place dubbed Glenelg, the first of many on its journey to the foothills of Mount Sharp. By mid-October scoop sampling had begun, with CheMin processing starting on Sol 70, 16 October, followed by SAM sample processing on Sol 93. By mid-December Curiosity was heading for a place called Yellowknife Bay and more contact science with a run of the SAM instruments. By 20 December Curiosity had reached Yellowknife and was entering a zone of different terrain that brought its own characteristics to rover handling. It was here that the first drilling operations would take place.

Designed for operations lasting a full Martian year (almost two Earth years), Curiosity is likely to carry on exploring the floor of Gale Crater and the foothills of Mount Sharp for many more years. And all the while, another rover – Opportunity – is still conducting science it has been delivering since February 2004. As 2012 drew to a close NASA was given approval to plan for a second rover mission for launch in 2020, following an orbiter in 2014 and a lander designed to probe deep beneath the surface for launch in 2016. Known as Curiosity 2.0, the new rover mission for 2020 will be based on its highly successful namesake and use spares, components and equipment from the current mission. By early 2013, with Curiosity relentlessly homing in on Mount Sharp, scientists were already discussing the landing sites for Curiosity 2.0.

BELOW A view of the lower front and underbelly areas of Curiosity in a mosaic of nine images taken by the MAHLI camera on 9 September 2012, Sol 34 of the rover's life on Mars. *(JPL)*

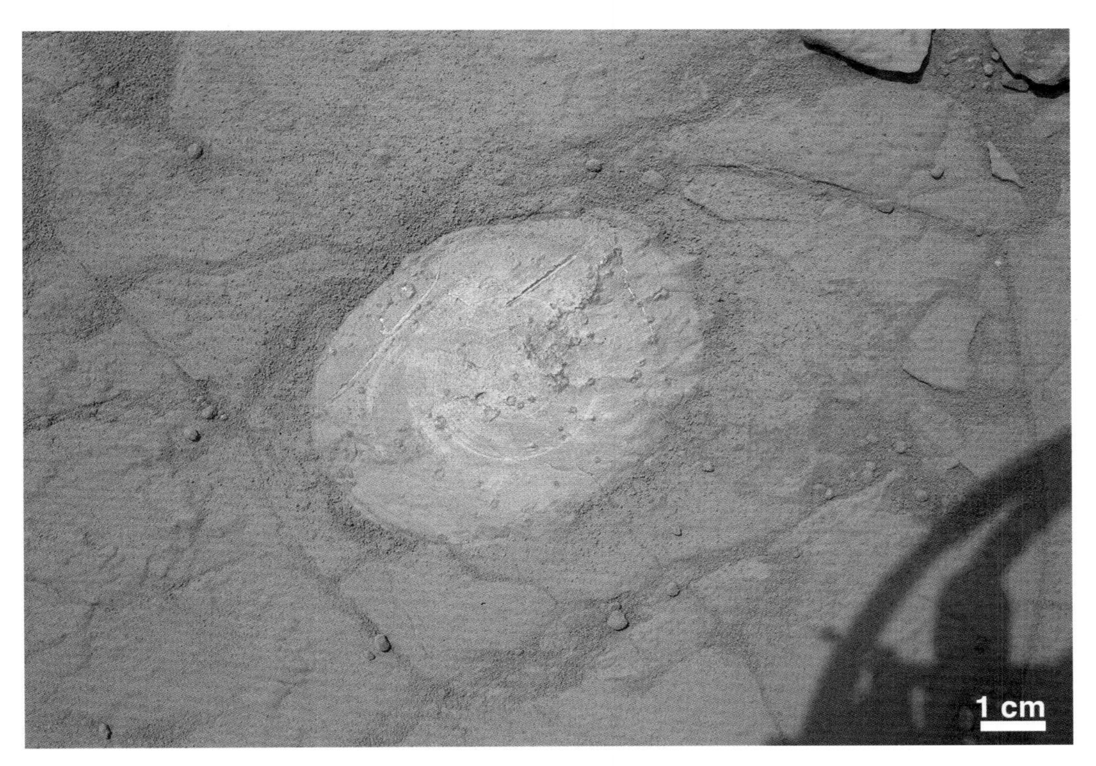

LEFT The MAHLI camera took this view of the first use of the Dust Removal Tool on 6 January 2013, Sol 150, from a distance of about 10in. The patch brushed away is about 1.8 x 2.4in across. *(JPL)*

BELOW A collage of different rocks sampled by NASA rovers on Mars, all built by JPL. The top two images show Spirit's observations at Gusev Crater, with a Viking landing site at lower left. A close-up of a soil target for Curiosity at Gale Crater is at lower right. *(JPL)*

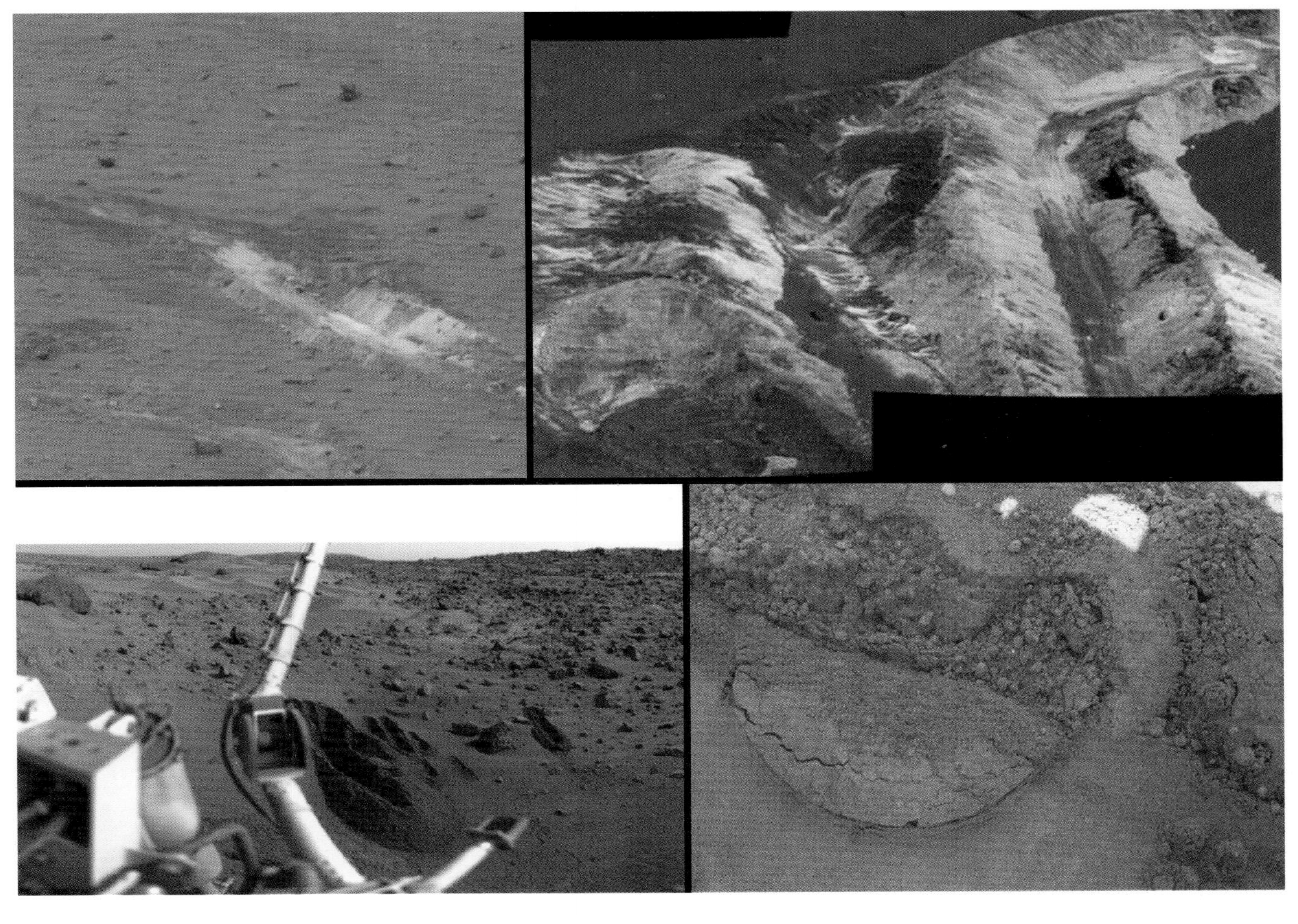

Postcards from Mars

OPPOSITE Curiosity takes images that together form a seamless mosaic in a 'self-portrait' at a site called Rocknest. *(JPL)*

RIGHT The surface of Gale Crater displays abundant evidence of a watery past early in the geological history of Mars. *(JPL/Caltech/Ken Kremer/Marco Di Lorenzo)*

BELOW Material that could not possibly have been blown by wind is strong evidence of fast-flowing water, at an outcrop dubbed Shaler. *(JPL)*

ABOVE Curiosity gets within arm's reach of a plain known as Snake River, a sinuous trail of dark rock cutting through lighter rock and sand dunes in this mosaic of images taken by Curiosity in early January 2013. *(JPL/Caltech/Ken Kremer/Marco Di Lorenzo)*

BELOW Yellowknife Bay, where Curiosity performed its first drilling operation on 8 February 2013. *(JPL/Caltech/Ken Kremer/Marco Di Lorenzo)*

ABOVE John Klein outcrop provides numerous examples of ideal sampling opportunities, but the pressure is on to get Curiosity moving toward its main goal – Mount Sharp. *(JPL/Caltech/Ken Kremer/Marco Di Lorenzo)*

RIGHT Curiosity takes a look back at its rover tracks. Note the prominent cooling fins on the nuclear power source. *(JPL/Caltech/Ken Kremer/Marco Di Lorenzo)*

OPPOSITE TOP **The equivalent of NASA's Lunar Roving Vehicle used by Apollo astronauts may be the legacy of rovers like Curiosity, when humans tread the dust of Mars.** *(NASA-JSC)*

OPPOSITE BOTTOM **An astronaut abseils down a sheer cliff face in the giant canyonlands on Mars in a future where robots and humans work together.** *(NASA-JSC)*

LEFT **Where it all began. An astronaut touches the Sojourner micro-rover, on Mars since 1997.** *(NASA-JSC)*

BELOW **One of two Viking landers gets a visit from curious humans investigating the effects of weathering on Mars in a future paved by autonomous rovers.** *(NASA-JSC)*

Glossary and abbreviations

APXS Alpha Particle X-ray Spectrometer.
ARA Airbag Retraction Actuators.
ASI/MET Atmosphere Structure Instrument/ Meteorology.
Athena Mars rover design prior to the MER missions.
Atlas V 541 Launch vehicle for MSL.
ATLO Assembly, Test and Launch Operations.
bps Bits per second.
BUDL Bridle and Umbilical Descent Limiter (for MSL).
CCD Charged Couple Detector.
CDR Critical Design Review.
CFC Chlorofluorocarbons.
cg Centre of gravity.
ChemCam Chemistry and Camera instrument (for MSL).
CheMin Chemistry and Mineralogy instrument (for MSL).
CHIMRA Collection and Handling for In-Situ Martian Rock Analysis (for MSL).
CHRS Cruise HRS (for MSL).
CMA Camera Mast Assembly (MER rovers).
CNES Centre Nationale d'Etudes Spatial.
Collimator Light focused by a slit into a parallel beam.
CO2 Carbon dioxide.
COTS Commercial Off The Shelf.
DAN Dynamic Albedo of Neutrons (for MSL).
dB Decibels.
DC Direct current.
DEA Descent Electronics Assembly.
DGB Disk-Gap-Band (parachute design).
DIMES Descent Image Motion Estimation Subsystem.
DRAM Dynamic Random Access Memory.
DRL Descent Rate Limiter.
DRT Dust Removal Tool (for MSL).
DSIF Deep Space Instrumentation Facility.
DSN Deep Space Network.
E Time to (+) or from (–) entry interface.
EADS European Aeroneutronic Defence and Space company.
EDL Entry, Descent and Landing.
EERPROM Electrical Erasable Programmable Read Only Memory.
ETPC Entry Terminal Point Controller (for MSL).
g Measure of gravitational force.
GaAs Gallium Arsenide.
GaInP Gallium Indium Phosphorous.
GB Gigabytes.
GC Gas Chromatograph (for MSL).
Ge Germanium.
GHz Gigahertz (billions of cycles per second).
GMT Greenwich Mean Time.
GN&CS Guidance Navigation & Control System.
GPHS General Purpose Heat Source (for MSL).
GSE Ground Support Equipment.
Hazcam Hazard avoidance camera.
HGA High-Gain Antenna.
HPA Heat Pump Assembly (MER rovers).
HRS Heat Rejection System (MER rovers).
HX Heat Exchanger (for MSL).
ICER Wavelet-based image compression format (for MSL).
IDD Instrument Deployment Device (MER rovers).
IMP Imager for Mars Pathfinder.
IMU Inertial Measurement Unit.
ISIL In-Situ Instruments Laboratory at JPL.
IVP Inertial Vector Propagation (for MSL).
j/m^2 Joules per square metre.
JPL Jet Propulsion Laboratory.
kN/m^2 Kilonewtons per metre squared.
ksi A unit of pressure equal to 1,000psi.
kW Kilowatt.
L The time of landing.
LC 17 Launch Complex 17.
L/D Lift over drag.
LED Light-emitting diode.
LGA Low-Gain Antenna.
LIBS Laser Induced Breakdown Spectrometer (for MSL).
LOCO Low Complexity Lossless Compression software (for MSL).
LPA Lander Petal Actuator.
m Metres (distance).
Mach The speed of sound.
MAHLI Mars Hand Lens Imager (for MSL).
MARDI Mars Descent Imager (for MSL).
MastCam Mast mounted camera.
Mastcam 34 MastCam with 34mm lens.
Mastcam 100 MastCam camera with 100mm lens.
MCO Mars Climate Orbiter.

MDI Mars Descent Imager (for Phoenix).
MEDLI MSL EDL Instrumentation (for MSL).
MER Mars Exploration Rover.
MER-1 Opportunity.
MER-2 Spirit.
M-FSK Multiple Frequency Shift Key (MER rovers).
MGP Mars Global Pathfinder.
MGR Mars Geological Rover.
MGS Mars Global Surveyor.
MHz Megahertz (millions of cycles per second).
Mini-TES Miniature Thermal Emissions Spectrometer.
MISP MEADS Integrated Sensor Plugs (for MSL).
Mission A Flown by Spirit rover (MER-2).
Mission B Flown by Opportunity (MER-1).
MLE Mars Lander Engines (for MSL).
MMP Mars Mobile Pathfinder.
MMRTG Multi-Mission Radioisotope Thermoelectric Generator (for MSL).
Monopole A single-pole antenna.
MPFL Mechanically Pumped Fluid Loop (for MSL).
mph Miles per hour.
MPL Mars Polar Lander.
MR-80 Descent engines for Viking.
MR-80B Descent engines for MSL.
MRO Mars Reconnaissance Orbiter.
MSL Mars Science Laboratory.
MSO Mars Science Orbiter.
MSSS Malin Space Science Systems.
mW Milliwatt.
N Newtons force.
N/m^2 Newtons per square metre.
NASA National Aeronautics & Space Administration.
Navcam Navigation camera (for MSL).
ORT Operational Readiness Test.
Pa Pascals (units of dynamic pressure).
PADS Powder Acquisition Drill System (for MSL).
Pancam Panoramic camera (on MER rovers).
Paschen Voltage necessary to arc across a gap according to pressure.
PDT Pacific Daylight Time.
PDV Powered Descent Vehicle (for MSL).
Phoenix Mars lander.
PHSF Payload Hazardous Servicing Facility located at Cape Canaveral.
PICA Phenolic Impregnated Carbon Ablator.
PMA Pancam Mast Assembly.
psi Pounds per square inch.
QMS Quadrupole Mass Spectrometer (for MSL).
RA Robotic Arm.
RAD Radiation Assessment Detector (for MSL).
RAD Rocket Assisted Deceleration.
RAM Random Access Memory.
RAT Rock Abrasion Tool (on MER rovers).
RCA Radio Corporation of America.
RCE Rover Computing Element (for MSL).
RD-180 Russian-designed first stage engine for Atlas V 541.
RED Rover Equipment Deck.
REM Rover Electronics Module (MER rovers).
REMS Rover Environmental Monitoring Station (for MS).
RF Radio frequency.
RHF Rover HRS (for MSL).
RHU Radioisotope Heater Units.
RMI Remote Micro Imager (for MSL).
rpm Revolutions per minute.
RSM Remote Sensor Mast (for MSL).
RTG Radioisotope Thermoelectric Generator.
SA/SPaH Sample Acquisition/Sample Processing and Handling (for MSL).
SAM Sample Analysis at Mars (for MSL).
S-band frequencies in the 2–4GHz band.
SI Standard Initiator.
SMSA Surface Mission Support Area.
Sol A Martian day of 12hr 40min.
SOWG Science Operations Working Group at JPL.
Soyuz-FB-Fregat Russian launch vehicle.
SSE Sensor Support Equipment (for MSL).
Suncam Tiny camera for photographing the Sun.
TCM Trajectory Correction Manoeuvre.
TDS Terminal Descent Sensor (for MSL).
TES Thermal Emissions Spectrometer.
TIRS Transverse Impulse Rocket System (MER rovers).
TLS Tunable Laser Spectrometer (for MSL).
Type I A flight path of less than 180° around the Sun.
Type II A flight path of more than 180° around the Sun.
UHF Ultra High Frequency.
UTC Universal Time Clock (formerly GMT).
VHF Very High Frequency.
W Watts of electrical energy.
W/cm^2 Watts per centimetre squared.
WEB Warm Electronics Box.
X axis Longitudinal (roll).
X-band Frequencies in the 7–11.2GHz band.
Y axis Lateral (pitch).
Z axis Vertical (spin).

Index

People